POLICYMAKING FOR CRITICAL INFRASTRUCTURE

Policymaking for Critical Infrastructure

A Case Study on Strategic Interventions in Public Safety Telecommunications

GORDON A. GOW
London School of Economics and Political Science

ASHGATE

Published by
Ashgate Publishing Limited
Gower House
Croft Road
Aldershot
Hampshire GU11 3HR
England

Ashgate Publishing Company
Suite 420
101 Cherry Street
Burlington, VT 05401-4405
USA

Ashgate website: http://www.ashgate.com

British Library Cataloguing in Publication Data
Gow, Gordon A.
    Policymaking for critical infrastructure : a case study on
    strategic interventions in public safety telecommunications
    1.Emergency communication systems - Security measures
    2.Public key infrastructure (Computer security) 3.Wireless
    communication systems - Security measures
    I.Title
    384'.043

Library of Congress Cataloging-in-Publication Data
Gow, Gordon A.
    Policymaking for critical infrastructure : a case study on strategic interventions
in public safety telecommunications / by Gordon A. Gow.
        p. cm.
    Includes bibliographical references and index.
    ISBN 0-7546-4345-X
    1. Telecommunication--Security measures. 2. Public key infrastructure (Computer
security) 3. Wireless communication systems--Security measures. I. Title.
    TK5102.85.G69 2005
    384'.043--dc22

                                                                    2004025399

ISBN 0 7546 4345 X

Printed and bound by Athenaeum Press, Ltd.,
Gateshead, Tyne & Wear.

Contents

List of Figures

List of Tables

Acknowledgments

I wish to thank Professor Robin Mansell for the kind words of encouragement that led me to pursue this book project. I am especially grateful to Peter Anderson, David Mitchell, Richard Smith and Roman Onufrijchuk for their enduring support over the years.

Preparation of the final manuscript would not have been possible without assistance from the Department of Media and Communications at the London School of Economics. Johanne Provençal's copyediting has no doubt improved upon the flow and clarity of my ideas.

The case study research presented in the book was funded through programs sponsored by Public Safety and Emergency Preparedness Canada (formerly Office of Critical Infrastructure Protection and Emergency Preparedness) and the Social Sciences and Humanities Research Council of Canada.

For taking that courageous decision to join me in London, this book is dedicated to Sulya Anne.

I am, of course, responsible for reporting on and interpreting the evidence included in this study. Any errors or omissions are entirely my responsibility.

List of Abbreviations

AEAA	Alberta E9-1-1 Advisory Association
ALI	Automatic Location Identification
ANI	Automatic Number Identification
ANT	Actor Network Theory
BCSPA	British Columbia 9-1-1 Service Providers Association
CAS	Call-path Associated Signalling
CISC	CRTC Interconnection Steering Committee
CLEC	Competitive Local Exchange Carrier
CRTC	Canadian Radio-television and Telecommunications Commission
CTA	Constructive Technology Assessment
CWTA	Canadian Wireless Telecommunications Association
EPC	Emergency Preparedness Canada
ESRD	Emergency Services Routing Digits
ESWG	CISC Emergency Services Working Group
FCC	Federal Communications Commission
FEMA	Federal Emergency Management Agency
ICLR	Institute for Catastrophic Loss Reduction
ILEC	Incumbent Local Exchange Carrier
LTS	Large Technical Systems
MIN	Mobile Identification Number
MSAG	Master Street Address Guide
NCAS	Non Call-path Associated Signalling
NENA	National Emergency Numbering Association
OAB	Ontario 9-1-1 Advisory Board
OCIPEP	Office of Critical Infrastructure Protection and Emergency Preparedness
PSAP	Public Safety Answering Point
PSEPC	Public Safety and Emergency Preparedness Canada
PSTN	Public Switched Telephone Network
SCOT	Social Construction of Technology
TIF	Task Information Form
VAS	Value-added Services
WEWG	CWTA Wireless E9-1-1 Working Group
WSP	Wireless Service Provider

Introduction

This book represents an effort to weave together developments in the field of disaster management with an approach known as 'Constructive Technology Assessment' to provide a new perspective on policymaking for critical infrastructure. The essential argument that I put forward is that well-founded public policy must be based on an understanding of the social roots of risk and vulnerability in large technical systems and that this understanding must come from studying how these systems grow and change as *socio-technical* entities. It is precisely this knowledge that is prerequisite for the design and adoption of appropriate public policy interventions to achieve sustainable societies nationally and internationally.

The public information infrastructure plays a vital role and is at the heart of all critical systems in advanced industrial societies, and as such, it provides the central point of focus in the book. On the one hand, regulatory reform has resulted in a tremendous wave of technological innovation and ground-breaking opportunities for enhanced value-added services that could reduce risk and vulnerability by providing continuity services to a wide range of businesses and government institutions. On the other hand, such developments harbour hidden risks, particularly in light of rapid technological change and competitive pressures among service providers seeking to retain control over strategic network elements.

In many respects, this book is a product of my professional interest in an understanding of the process of growth and change in the public information infrastructure as a necessary step toward the identification and assessment of public policy intervention strategies. Originally, the project began with a general inquiry into the role of the telecommunications sector within Canada's National Disaster Mitigation Strategy, a public-private joint initiative launched in 2000/2001 and intended to reform that country's policy framework for emergency management. At the same time, I also discovered that Canada was not alone in adopting this mitigation-oriented policy framework but that such a framework was being introduced in countries around the world, corresponding in part to the aims and spirit of the United Nation's International Decade for Natural Disaster Reduction that had run from 1990 to 1999. Mitigation, it seemed to me, had emerged as the new *paradigm* for emergency and disaster management around the world.

As I continued my research, two important research questions surfaced. First, what would be the impact of a national mitigation strategy on the information infrastructure? Second, and conversely, how might the public information infrastructure be drawn upon to actively promote the objectives of such an initiative? Working on these questions, I began to realize that the term 'mitigation' itself was problematic and that many of the existing programs in emergency telecommunications, business continuity planning, and critical infrastructure protection did not seem to conform to what I sensed to be the more fundamental concerns implied by Canada's National Disaster Mitigation Strategy or the promising objectives of other national and international initiatives.

This realization led me to conclude that the term 'mitigation,' despite the volumes that have been written on it, remains an ambitious but *ambiguous* idea that must be more clearly defined so it can stand apart from—and yet remain integrated with—other activities in disaster management. To achieve clarity on this matter, I looked beyond the typical descriptive model of disaster management based on a four-phase cycle and adopted an explanatory model of disasters that could more clearly account for the root causes of vulnerability and unsafe conditions in society. My adaptation and interpretation of the 'Pressure and Release' model for the management of critical infrastructure (found in the first chapter) suggests that mitigation-oriented policy research is appropriately directed toward the early processes and influences on network development—that is to say, the social roots of risk and vulnerability.

Having worked out a more satisfactory definition of mitigation, using what I felt to be a more suitable model for policy research, I then turned my attention to the theoretical and methodological issues of understanding and studying the fundamental forces of growth and change in large technical systems. This stage of my research led me to look for a body of theory that was consonant with an application of the mitigation paradigm, yet suitable to a study of critical infrastructure. The result is an assemblage of several related approaches from science and technology studies and Constructive Technology Assessment, operationalized through a selection of telecom policy research literature. Together these sources are integrated and form an analytical framework to study growth and change in networked systems, providing what I believe is a much-needed bridge between a growing body of scholarly research in science and technology studies, and the practical and pressing concerns of policymakers working in disaster mitigation and critical infrastructure protection.

Empirically, this study is intended to make a comparatively modest contribution to the history of technology and to applied policy research. The study also presents an operationalization of the analytical framework used to conduct a detailed analysis of the development of one instance of growth and change in the public information infrastructure. Historically, the study provides a detailed account of the development of a public safety telecommunications service known as Wireless 'enhanced' 9-1-1 ('Wireless E9-1-1'), a location-based emergency service for mobile telephone customers currently being deployed in North America. For applied policy research, the final chapter sets out a number of suggestions and examples of alternative intervention strategies based on observations made in the case study. These suggestions may prove useful to policymakers and other analysts who are considering how national mitigation strategies impact public policy for critical infrastructure and perhaps more importantly, how careful management of these large technical systems may serve to support the wider social objectives of mitigation-oriented initiatives.

Chapter One, 'A Public Policy Challenge,' introduces the background, rationale, and focus for the book. In this chapter I argue that modern societies face an interdependency dilemma that remains largely unaddressed in public policy that purports to be mitigation-oriented. I then critique and rethink the concept of mitigation by introducing the Pressure and Release model. I discuss the implications this model has for the management of critical infrastructure in general and for the study of the public information infrastructure in particular.

Chapter Two, 'The Design Nexus,' presents the theoretical and methodological foundations that I adopt to study growth and change in critical infrastructure. Here I look to Constructive Technology Assessment as a primary approach to the study of growth and change because of its emphasis on the early design phase of technology development. I also draw on Actor Network Theory and Social Construction of Technology to provide a theoretical model of technology dynamics for the book. I then bring together elements from each to establish a basic methodology and analytical framework for studying technological change. This framework takes the form of an 'intervention matrix' that serves as the principle analytical tool used in the case study on public safety telecommunications and throughout the book.

Chapter Three, 'Turning to the Empirical,' draws on research undertaken on Large Technical Systems (LTS) to establish the operational parameters needed to conduct empirical research on technological change. Among these parameters, I address the problem of setting boundaries on networked infrastructure and I identify 'interconnection' as a core operational concept. Interconnection also provides a crucial third dimension to the intervention matrix. On the basis of a selected range of literature in telecom policy research, I identify and discuss various issues related to interconnection, including technical standardization and the political economy of network design. Later in the book (Chapter Six), I return to the literature on Large Technical Systems and Social Construction of Technology to look at other issues associated with stakeholder participation in the management of critical infrastructure.

Chapters Four, Five, and Six present the case study in public safety telecommunications. The case follows in considerable detail the development and deployment of 'Wireless E9-1-1' in Canada. Wireless E9-1-1 is a public safety innovation that provides location-enhancement for mobile phone customers who dial emergency services. It is now being deployed in North America, as well as in Europe and Australia where it is variously referred to as 'E1-1-2' or 'MoLI' (Mobile Location Information). The service has significant impact at multiple levels of public information infrastructure, including service reconfiguration, operational procedure, and investment in new technology, and it provides a glimpse of the complexity of technological change in modern critical infrastructure.

Each of the three case study chapters is dedicated to one 'facet' of the Wireless E9-1-1 story. Chapter Four, 'The Standardization Effort,' focuses on the adoption of technical standards for Wireless E9-1-1 and illustrates the wide range of considerations that influence technological change in modern infrastructure and the corresponding challenges of identifying appropriate regulatory intervention to ensure public interest objectives. Chapter Five, 'Innovation and Experimentation,' looks at the importance of industry-driven efforts toward innovation and experimentation in new value-added services and considers the role of policymakers in identifying obstacles to such efforts among new entrants and third parties. Chapter Six, 'Communities of Experts,' takes up the third facet of the case study and examines the difficulties associated with an expanded range and diversity of stakeholder participation in the management of critical infrastructure.

Finally, Chapter Seven, 'The Structures of Intervention,' provides an overall analysis of the Wireless E9-1-1 case study. This is undertaken in three steps. First I present a 'socio-technical mapping' of key actors and issues in the case and I discuss several predominant intervention strategies that became evident during my investigation. Second, I examine the case as a single unit of analysis using the intervention matrix to portray an overall structure to the observed interventions. I then critique this apparent structure to provide suggestions for its improvement, supported by real-world examples that together provide an alternative structure of interventions that I believe more closely corresponds to the principles and objectives of the mitigation-oriented policy framework presented at the beginning of the book.

The book covers a lot of ground and some readers may be surprised at the unusual perspective it brings to the study of critical infrastructure. My hope is that readers will nonetheless find the book fruitful to the extent that it presents some new ideas for this important subject of study and encourages further theoretical and empirical work along the lines I have introduced. For practitioners and policymakers, I hope the book lends a measure of clarity to the problematic notion of 'disaster mitigation' and provides some practical insight for the design of programs and policy interventions. Scholars in science and technology studies may simply be interested in the historical details of the Wireless E9-1-1 case or my adaptation of SCOT, ANT, and LTS to a field of applied policy research. Some readers will find the details of the case study rather profuse; however, I feel it has historical value, so I have tried to include as much detail as possible even where such detail may not be directly related to the book's primary aim.

Chapter 1

A Public Policy Challenge

Creativity involves breaking out of established patterns in order to look at
things in a different way.

-Edward de Bono

The Interdependency Dilemma

The risk of local accidents escalating rapidly into nationwide critical incidents is
perhaps nowhere more acute than in advanced industrial societies, where
telecommunications continue to foster a complex integration of social institutions
with large technological systems. In a series of revisions to its 'Guidelines for the
Security of Information Systems' first published in 1992, the Organisation for
Economic Co-operation and Development (OECD) recognized the growing state of
infrastructure interdependency, observing that 'ever more powerful personal
computers, converging technologies and the widespread use of the internet have
replaced what were modest, stand-alone systems in predominantly closed
networks' (Organisation for Economic Co-operation and Development, 2002).

The interdependency dilemma on the one hand refers to the tight coupling of
modern institutions through telecommunications. The terrorist attacks on the
World Trade Center towers in New York City not only caused a tragic loss of life
and property but revealed the vulnerability of North America's communications
infrastructure. In the aftermath of the incident, reports in the trade press indicated
that a major U.S. telecommunications carrier, Verizon, had five central offices
serving some 500,000 telephone lines in the vicinity of the World Trade Center and
that more than six million private circuits and data lines passed through switching
centres in or near the site of the collapsed twin towers. Additional reports claimed
that the AT&T and Sprint switching centres in the World Trade Center were
destroyed, as were numerous cellular base stations in the vicinity (Angus
Telemanagement Group, 2001). Data networks for major corporations, including
AOL Time Warner and other broadband services, were also disrupted by the
collapse of the towers (Ray, 2001). A society increasingly bound together by
electronic networks is at constant risk of telecommunications outages ascending to
major incidents with consequences that extend far beyond the initial site of impact.
The terrorist attacks of 9/11 were not about perpetrating a telecom outage *per se*,
yet they underscore the vulnerability of the modern telecommunications
infrastructure to unforeseen events.

On the other hand, the interdependency dilemma also refers to the fact that minor, seemingly innocuous, failures in the telecommunications network may have social effects that are far-reaching and often surprising in their immediacy. In the summer of 1999, for instance, a technician working at a telephone switching office in downtown Toronto accidentally dropped a tool into a power supply unit causing a minor electrical fire. The next morning, the story was splashed across the Canadian media. What started as a minor incident had erupted into a nationwide 'phone crisis,' for not only did the Toronto switch fire affect local voice telephone service, its impact rippled outward across the city and quickly spread to affect businesses from coast to coast. As it happened, the switching office turned out to be a key node in a nationwide data network that supports transaction services including credit card authorization and automated teller machines. Locally, the disruption spread beyond plain old telephone service to affect data services for numerous private and public institutions, including those with a vital social support function. Details of the incident were recorded in extensive press coverage of the event (Blackwell, Craig and Bell, 1999; Cheney, 1999):

- 113,000 telephone landlines were disrupted
- mobile phone service was disrupted
- Art Gallery of Ontario security system went offline
- travel agents went offline (one location lost some $30,000 worth of sales)
- some retailers were unable to process credit/debit card transactions
- Ontario Lottery terminals went offline
- some stock trading was affected
- law offices were unable to close real estate deals (clearing house for title searches was offline)
- hundreds of bank branches went offline
- almost 1/10 of cash machines across Canada were out of service for parts of the day
- sequencing for about 570 traffic lights was disrupted
- Toronto police telephone and computer systems were disrupted
- Hospital for Sick Children and several others were affected by disruption to pager and telephone services
- poison control centre and medical information hotlines were affected
- 9-1-1 system was maintained but capacity to handle calls was impaired.

A similar incident occurred in November 2000, when a contractor in a Chicago-area rail yard accidentally cut a cable providing access to the Canadian Venture Exchange (CDNX) trading system. The accident subsequently interfered with trading activities right across Canada when backup provisions for the system reportedly failed to work (Cattaneo, 2000). Modern telecommunications systems foster interdependencies whereby a distant event in a seemingly innocuous location can lead to serious disruptions for business and community organizations across vast distances. Some other examples of the interdependency dilemma (Robinson, Woodard and Varnado, 1998):

- The satellite malfunction of May 1998. A communications satellite lost track of Earth and cut off service to nearly 90 percent of the nation's approximately 45 million pagers, which not only affected ordinary business transactions but also physicians, law enforcement officials, and others who provide vital services. It took nearly a week to restore the system.
- The Northridge, California, earthquake of January 1994 affecting Los Angeles. First-response emergency personnel were unable to communicate effectively because private citizens were using cell phones so extensively that they paralyzed emergency communications.
- Two major failures of AT&T communications systems in New York in 1991. The first, in January, created numerous problems, including airline flight delays of several hours, and was caused by a severed high-capacity telephone cable. The second, in September, disrupted long distance calls, caused financial markets to close and planes to be grounded, and was caused by a faulty communications switch.

Increasing interdependency means that minor, local disruptions in telecommunications service due to a major earthquake, severe weather incident, workplace accident, or terrorist attack could quickly propagate nation-wide and even continent-wide second-order incidents of potentially disastrous magnitude in terms of unexpected social, economic, and even environmental repercussions. Such high stakes scenarios have not gone unnoticed by policymakers, particularly in the wake of the Y2K crisis and the terrorist attacks of 11 September 2001. These events have prompted a growing recognition of the interdependencies fostered by modern telecommunications, and serious efforts at reducing so-called 'common mode failures' (Hellström, 2003, p. 376) are now evident within the changing context of policy research for critical infrastructure (Gheorghe, 2004). The operational side of emergency preparedness and national security has also taken up the challenge. In February 2003, for example, the United States government issued its *National Strategy for the Physical Protection of Critical Infrastructures and Key Assets*. In this strategy it calls for risk management based on the coordination of public and private interests:

> Because of growing interdependencies among the various critical infrastructures, a direct or indirect attack on any of them could result in cascading effects across the others. ... Critical infrastructures rely upon a secure and robust telecommunications infrastructure. Redundancy within the infrastructure is critical to ensure that single points of failure in one infrastructure will not adversely impact others. It is vital that government and industry work together to characterize the state of diversity in the telecommunications infrastructure. (United States, 2003, p. 49)

Similarly, in Canada the federal department of Public Safety and Emergency Preparedness Canada (PSEPC) has recognized the wide social interdependencies linked to the modern public telecommunications infrastructure:

> In this 'Information Age,' critical infrastructure is a key enabler to the modern economy. It is complex, interconnected and interdependent and relies heavily on information technology. Disruptions in one infrastructure could produce cascading disruptions across a number of other infrastructures, with significant economic and social consequences to Canada and Canadians. (Canada, 2001)

Concerns expressed in these official government statements are supported by research into critical infrastructure interdependencies. According to a study by Masera and Wilikens, this interdependency is a product of three major trends shared across the infrastructures that enable modern societies to function as they do. The authors summarize these as (1) the increasing complexity of technological systems; (2) a high degree of interconnectedness at both technical and organizational layers between these systems; and (3) a growing reliance on information and communication technologies to provide support for and control over various systems and subsystems. It is with this third trend that telecommunications enters into the picture as the vital link that integrates a variety of critical infrastructures into a complex network of co-dependent systems:

> The interconnections among infrastructures are facilitated by the internal use of ICT [Information and Communication Technologies], and by the availability of affordable means of establishing fast and reliable data communications. These communications means are beginning to be considered as a sort of public information infrastructure, and are increasingly employed for exchange of information and access to new value-added services such as trade tools for actors in a given market sector, and for getting connected with other infrastructures. ...
>
> ... the fact is that each and every infrastructure, and that means society at large, is attempting to benefit from the public information infrastructure. The end result is an overall, global dependence on a limited set of hardware and software technologies, with evolving and not yet mature interconnection and business models. (Masera and Wilikens, 2001)

An important facet of this observation is that the trajectory of the public information infrastructure remains contingent and relatively immature, reflecting the turmoil created by digital convergence and widespread regulatory reforms. On the one hand, such conditions breed greater vulnerability as a result of unforeseen risks and unexpected consequences of investment decisions. On the other hand, this contingency represents an opportunity for policymakers to embark on a coherent program of long-term risk reduction through the social shaping of critical infrastructure. It is in response to both the hidden risk and foreseeable opportunity that this book is principally addressed.

The astute reader may also have noticed a discrepancy between the 'public information infrastructure' mentioned above and the current telecommunications infrastructure that we tend to think about with respect to the regulation of circuit-based voice services. In their study, Masera and Wilikens are concerned with the information infrastructure made up of data networks that have in the past often been deployed separately from circuit-based voice networks. In other words, they are concerned with the important distinction to be made between two types of information services that were originally established as separate infrastructures. Yet, the circuit-based voice network tends to remains synonymous in many countries with *regulated public access telecommunications*. While we should be careful not to conflate the two infrastructures because of their unique functional qualities and regulatory status, this situation is changing rapidly as voice and data services converge at trunk-side and as new voice-over-IP (VoIP) services and other digital standards are deployed at line-side. As voice service continues to evolve into just another data application, regulated public access telecommunications will become increasingly difficult to distinguish, either analytically or operationally, from other data networks. Therefore even *within* regulated public access telecommunications, we find a growing trend toward convergence of systems and new forms of infrastructure interdependency.

An integrated public information infrastructure may be nascent but even today we need not look far to glimpse the complexity of the emerging interdependencies. The electricity infrastructure is a case in point, caught in a double bind between supplying commercial power to operate communication networks and using communication networks to supply commercial power:

> An advanced management of the generation and distribution of electricity has been made possible by the establishment of monitoring and control systems relying on complex communications systems. In addition electric companies rely on information exchanges for their connections with the energy value chain and their customers. Thus *it is now possible to speak of the electricity infrastructure as composed of the power grid and an associated information network.* [emphasis added]
>
> ... energy producers and distributors, while trying to develop their markets and enhance their efficiency, are getting [*sic*] strongly dependent on the public open Information Infrastructure on both the demand and supply sides. They have to communicate with the other energy market actors, as well as complete the energy offer with information-related services. These could be energy information services related to the energy loads, generation and consumption, or pricing and billing; or derived services that take advantage of the link established with residential or industrial customers (for instance, alarm management, heating, etc.). (Masera and Wilikens, 2001)

From this example one can begin to further grasp how the public information infrastructure fosters wider interdependencies that are foundational for a range of services stretching across geographic, social, and economic strata. A major challenge for traditional approaches to the management of critical infrastructure, however, is that public policy and the resulting publicly funded programs for emergency management have encountered difficulties in coming to grips with these profound socio-economic interdependencies either as they exist in the present or as they are likely to intensify in the future. Not coincidentally, this difficulty reflects a much larger concern with the management of risk in modern societies, where there has been an increasing recognition of the need to reconcile technological development with wider social policy objectives such as privacy, economic and environmental sustainability (Misa, Brey and Feenberg, 2003).

I wish to offer two possible explanations for this present difficulty in the management of critical infrastructure, both of which will be taken up further in this and other chapters. The first explanation is linked to the dominant conceptual model used in emergency management, which is descriptive as opposed to explanatory and which is best applied to units of analysis at the operational level rather than those more appropriate for the analysis of long-term public policy concerns. The second explanation is linked to inherent problems in coordinating action across the various domains of stakeholders that must be involved if the interdependency dilemma is to be effectively managed over the long-term. This problem of stakeholder coordination is, in the first instance, one of identifying and adopting a coherent public policy framework for technology assessment aimed at fostering a culture of what Thomas Hellström has termed 'responsible innovation' (Hellström, 2003).

Despite the difficulty of these challenges, there is nevertheless evidence to indicate that important progress is being made on both fronts. Following the conclusion of the United Nations International Decade for Natural Disaster Reduction (IDNDR) at the end of the 1990s, national governments around the world have begun to recognize the social and economic continuities that exist between technological innovation and risk management, as well as the close affinity both have with the sustainable development movement. At the same time, the dominant conceptual model employed in the field of emergency management has come under criticism for providing an inadequate theoretical framework for long-term risk reduction strategies. In the following sections, I will examine the trend toward public policy thinking based on long-term risk reduction, its guiding principles and aims for disaster management and the critical challenge it presents to the dominant conceptual model employed in emergency management. This discussion provides necessary background against which to identify gaps in current policy research concerning the interdependency dilemma, the public information infrastructure, and the management of critical infrastructure.

Shifting to Mitigation in Emergency Management

The1990s were designated by the United Nations as the International Decade for Natural Disaster Reduction (IDNDR), the central aim of which was, among other things, 'to reduce the loss of life, property damage, and social and economic disruption caused by natural disasters …' (Jeggle, 1999). Despite criticisms that claim the IDNDR focussed too much attention on natural hazards at the expense of dealing with the root social causes of vulnerability (see Blaikie *et al.*, 1994), it did nonetheless enrich international debate about natural hazards and vulnerability, making important connections in policy terms between disaster management and sustainable development (Bruce, 1999). For one thing, the IDNDR encouraged a new approach to public policy strategies based on the long-range reduction of risk and vulnerability, also known as *disaster mitigation*, and raised awareness of the important role that it could play in enhancing the quality of life for both the developing and developed regions of the world.

With the conclusion of the IDNDR in 2000, the UN then launched the International Strategy for Disaster Reduction (ISDR) in order to implement the ideas and program concepts established in the previous decade and to provide a backdrop for the emergence of national disaster mitigation strategies in countries around the world. Much of the ISDR's work is focussed on developing regions of the world such as Africa, Latin America, the Caribbean, and Asia, assisting in the development of disaster mitigation programs in those regions most prone to suffer socially and economically from the impact of natural hazards. The IDNDR also served to push mitigation on to the agenda in advanced industrial nations, however, where rapidly rising costs of disaster recovery are cause for alarm among both private and public institutions. In 2001, for instance, the Government of Canada in cooperation with the Insurance Bureau of Canada announced a public consultation on the formation of a National Disaster Mitigation Strategy (Office of Critical Infrastructure Protection and Emergency Preparedness, 2001). By contrast, the United States Federal Emergency Management Agency (FEMA) has been promoting mitigation since the mid-1990s as a measure to reduce the costs to government in assisting citizens to recover from losses incurred as a result of natural disasters (Federal Emergency Management Agency, 1996). Most recently, the Australian government launched its National Disaster Mitigation Programme, noting that,

[P]riority should be given to projects that are derived from or contribute to strategies to address the fundamental causes, rather than symptoms, of Australia's natural disaster related problems and that brings long-term natural disaster mitigation benefits and, in addition, environmental, economic, and social benefits. (Australian Government, 2003, p. 4)

This fundamental principle common to all disaster mitigation strategies emphasizes the necessity of public policy measures to extend beyond a response to threats posed by natural hazards and to engage with the *underlying causes* that lead to the formation of risk and vulnerability in society in the first place. Furthermore, the principle introduces the claim that disaster reduction efforts might work in a coordinated way to support other social policy objectives, thereby expanding significantly the range of stakeholders with a potential interest in such programs.

I take this appearance of a policy shift toward mitigation-oriented thinking as evidence of a fundamental transformation in approach, due in part to an increased awareness of the wide societal value of adopting long range, integrated risk management and to a growing sensitivity to the cost, both social and economic, of natural disasters. Casual observers might think that there is an obvious connection between emergency management and long-range risk reduction, or sustainable development. Yet, it is the contention of those who advocate the development of national mitigation strategies that current practices and programs in emergency management may be necessary but that they are not sufficient conditions to contribute effectively to this recent shift in public policy thinking. As I noted above, part of this problem lies with the dominant model of emergency management within which the idea of mitigation is typically understood. The other part of the problem is the framework for coordinating stakeholder interests across diverse social and economic domains. While policy thinking continues to be hobbled by an inadequate and ambiguous concept of mitigation, emerging initiatives in the form of national and international mitigation strategies clearly represent an effort to establish a new policy framework for coordinating actions across social and economic domains of activity with the aim of fostering long-term risk reduction.

By way of illustration, I will now turn to describe briefly the intent and key principles embodied in one example of a national mitigation strategy. Let me acknowledge at the outset that the Canadian strategy is not intended to represent an exemplary strategy (especially given the extent of commitment and dedication to mitigation programs found in the United States and other countries) but serves to convey in a relatively straightforward manner many of the principle characteristics informing this recent shift in public policy thinking, while providing the context for a case study on public safety telecommunications.

Canada's National Disaster Mitigation Strategy

Within the wider context of the IDNDR and perhaps inspired in part by mitigation initiatives in the United States, Canada's overall approach to disaster management was scrutinized following what appeared at the time to be an unprecedented string of natural disasters between 1996 and 1998 that included the Saguenay River and Red River floods in 1996 and 1997 respectively, and the Eastern Canada ice storm in 1998. Alarmed at the growing cost of insurance payments arising from these severe weather events and spurred on by the looming earthquake threats on both the West and East coasts, the Insurance Bureau of Canada, in partnership with the University of Western Ontario and others, established the Institute for Catastrophic Loss Reduction (ICLR). The ICLR was to serve as a research body mandated to examine ways of mitigating the cost of natural disasters in Canada. In the autumn of 1998, the ICLR partnered with federal government department Emergency Preparedness Canada (EPC) to consider a national approach to reducing the impact of natural disasters.[1] The primary objective of this public-private partnership was to modify Canada's approach to disaster management by placing a greater emphasis on sustained efforts at long-term risk reduction. This partnership represents an important turn in Canadian public policy inasmuch as it marks a shift away from predominantly response-oriented emergency management toward a

---

[1] EPC was later changed to the Office of Critical Infrastructure Protection and Emergency Preparedness (OCIPEP), and again in 2003 to Public Safety and Emergency Preparedness Canada (PSEPC). This series of changes is partly a result of the upheaval and uncertainty in public policy thinking about disaster management and critical infrastructure protection that reflects a growing awareness of vulnerability stemming from at least three major influences: climate change, the proliferation of computer viruses, and fallout from the terrorist attacks of 11 September 2001.

proactive policy aimed at building what the ICLR termed 'resilient communities,' a term suggestive of initiatives associated with sustainable development rather than disaster management.

In due course, the Canadian government launched its National Disaster Mitigation Strategy following a series of workshops held across the country with a diverse group of community and professional stakeholders. Development of the national strategy was also aided indirectly by the findings of a parliamentary subcommittee that had held extensive hearings in 1999 to provide recommendations on reducing the impact of disasters on Canada's economy (Canada, 2000).

The initial motivation for a National Disaster Mitigation Strategy stemmed directly from the losses suffered as a result of a devastating ice storm that had struck eastern Canada in 1998 (Nicolet, 1999) and indirectly from international influences such as the United States and the UN's International Decade for Natural Disaster Reduction. In many respects, the events of the ice storm fuelled a discourse that had been simmering in Canada throughout the 1990s (Newton, 1997) and what emerged from the early workshops was a clear sense that a national mitigation policy is necessary if Canada is to avoid the burden of heavy social and economic losses caused by future severe weather events and earthquakes. As reported by the ICLR and others, the cost of weather-related disasters in Canada has risen sharply in the past two decades and this in conjunction with other factors is seen to be contributing to the increased vulnerability of Canadian society in a number of areas:

- Economic prosperity (i.e., greater prospects for losses to occur)
- Densification in urban centres
- Aging infrastructure
- Climate Change.

Potential future losses resulting from natural disasters not only include loss of life and destruction of property, but also and perhaps more significantly in terms of long run impacts include secondary impacts related to lost production and industrial competitiveness; restorative costs associated with disaster recovery; unanticipated personal hardship; and negative effects on public (and market) confidence. Central to mitigation-oriented public policy is the view that these kinds of losses can be significantly curtailed in advance through a sustained program of risk reduction. Above all, what we find in the Canadian discourse is that the key to a successful mitigation strategy is to design programs that will blend it into everyday practice as seamlessly as possible. Mitigation, as the Institute for Catastrophic Loss Reduction (ICLR) has pointed out, needs to be grounded in a set of cultural norms—a 'culture of mitigation'—that views risk reduction as an essential aspect of everyday life (Institute for Catastrophic Loss Reduction, 1998). Mitigation-oriented social policy is thus regarded as an enlightened approach to building and maintaining civil society, a view clearly reflected in several of the key themes that emerged from the early round of consultations that took place in 1998.

Among those key themes, several echoed the view that a mitigation strategy needs to be extended across a range of public policy undertakings. For instance, it became clear that participants in the consultation process felt that mitigation should be seen not as a one-time cost but rather, as an ongoing investment in long-term safety and economic security. Partnerships between government and the private sector are deemed necessary for such investment, and the local community is considered the most appropriate site for mitigation projects to be undertaken. A number of other principles surfaced during the consultations. Flexibility in methods was considered essential in order to address the wide variety of regional conditions across Canada. Education and awareness of risk along with applied research to identify and better understand risk were seen as equally important principles behind the culture of mitigation that would ultimately provide sustained support for a national strategy. Further was the assertion that 'mitigation measures must be compatible with other important public policy objectives' (Institute for Catastrophic Loss Reduction, 1998). For the telecommunications sector this assertion means that mitigation initiatives must attempt to work within the current policy framework as much as possible. This constraint presents both a challenge and an opportunity in the development of intervention strategies for the management of critical infrastructure and is a major consideration taken up in this book.

The ICLR/EPC consultations also produced a set of specific guidelines, or 'mechanisms for action' as they were termed in the report. Funding formulas formed the underpinning of these guidelines, essentially creating a framework for several types of support. Funding opportunities are considered in both pre-impact and post-impact stages of a disaster. Pre-impact funding would be guided by applied research to identify priorities within Canadian communities. Post-impact funding would be guided by assessments made during the recovery phase following a disaster and would be used to support ongoing mitigation projects linked to existing public works programs. The ICLR also recommended that Canada adopt a model similar to the United States and expand the current post-impact disaster financial assistance to prevent future reoccurrence through specific mitigation projects linked to rebuilding or recovery initiatives (Insurance Bureau of Canada, 1999).

The strategy would also require the private sector to become a major participant in mitigation programs and would seek ways to encourage the business community to initiate and fund mitigation projects in conjunction with public policy objectives. Coordination and leadership for the strategy would come from three sources: governments and the private sector committing resources toward mitigation; the creation of a national mitigation secretariat; and the formation of a national mitigation partnership.

The question remains as to how extensively the National Disaster Mitigation Strategy will inform public policy thinking in the management of critical infrastructure, in general and in telecommunications policy, in particular. In June 2001, the Government of Canada conducted a series of workshops on the development of a National Disaster Mitigation Strategy involving all levels of government, non-governmental stakeholders and the private sector (Office of

Critical Infrastructure Protection and Emergency Preparedness, 2001). Based on the outcome of these consultations, it is conceivable that the current national policy for emergencies will be revised to reflect, at least in part, the mitigation framework proposed by the ICLR. Recent developments stemming from the attacks of 11 September 2001 and growing threats related to cyberterrorism have led to a reassessment of priorities and the redirection of resources within governments around the world, sometimes taking resources away from budding mitigation programs. Nevertheless, governments remain committed to this new line of policy thinking, including the Canadian government, which issued a press release late in 2002 indicating that 'in cooperation with provincial-territorial partners and other stakeholders, [Canada] is continuing to advance the development of options and recommendations for a [National Disaster Mitigation Strategy] with a view to finalizing proposals during 2003' (Office of Critical Infrastructure Protection and Emergency Preparedness, 2002). Recent developments seem to have stalled, however, and Canada's office of Public Safety and Emergency Preparedness was criticized in a parliamentary report for not moving quickly enough on the NDMS (Canada, 2004, p. 16, 49). Moreover, the Australian government's recent launch of an extensive National Disaster Mitigation Programme for 2003-2004 (Australian Government, 2003) will likely add to the pressure on countries like Canada to continue with their efforts in this direction.

Implications for Telecommunications Policy and Regulation

If Canada and other countries continue to travel down this road of developing mitigation-oriented public policy, what implications might such policy hold for the management of critical infrastructure? Table 1.1 summarizes the original ICLR/EPC framework developed for Canada's National Disaster Mitigation Strategy and highlights a general set of key principles and guidelines that define similar initiatives in other countries, which establishes important parameters for this general approach to policy as it might apply to management of the public information infrastructure.

The first major objective is described as a 'sustained program of action' based on investment-oriented thinking and requiring a long-term commitment of resources. In terms of the public information infrastructure this means the creation of incentives for stakeholders to build mitigation programs into current practice. One conceivable way to do this is to add risk reduction as a regulatory consideration in the governance of Canada's telecommunications infrastructure. While this approach would not be unlike the ongoing investment in other public works programs, as with bridges or dams, it is an unconventional idea for a telecommunications sector undergoing extensive regulatory liberalization. Nevertheless, in terms of policy this approach could translate into a regulatory requirement to ensure that some form of publicly accountable risk assessment is incorporated into the maintenance, development, and deployment of telecommunications infrastructure and services. By introducing a procedure to guide management of critical infrastructure in this way, a sustained program of action is more likely to be feasibly maintained over time. Some further aspects of this idea are explored in subsequent chapters.

Table 1.1 Principles of Canada's National Disaster Mitigation Strategy

Major Objective	Associated Principle	Mechanism for Action
Sustained program of action	Investment-oriented thinking	Ongoing resource commitments
Consistency in method and application	Partnership-based approach	National mitigation partnership; standards
Expansion of expertise through interdisciplinary exchange	Flexibility and local orientation	Innovation and leadership in many fields
Dissemination of best practices	Promotion of education and awareness	National Mitigation Secretariat; industry forums
Framework for rational allocation of resources	Alignment with current policy frameworks	Policy review

The second objective of 'consistency in method and application' suggests that a national (and possibly international) approach is needed to ensure the development and application of coherent baseline standards to support mitigation-oriented activities. Flexibility and local orientation, however, will remain paramount considerations and a combination of consistency adapted to local conditions will depend on good working partnerships between government and the private sector. In this light, telecommunications policy and regulation might be applied to ensure that processes are designed to cultivate a level playing field through nationally recognized baseline requirements flexible enough to meet local needs. Perhaps most significantly, telecommunications policy instruments could be used to create a forum for the dissemination of best practices among users and service developers in order to enable partnerships and to support innovative ideas. In addition, selective intervention by the regulator may be necessary in certain cases where carriers and other service providers may be engaged in making critical path dependent decisions with implications for risk management. Finally, as stipulated in the original Canadian mitigation strategy, these objectives should align as closely as possible with the current telecommunications policy framework.

If a basic focus for a national mitigation strategy is indeed the underlying cause of risk and vulnerability in society and if we apply this focus to the above set of principles, then it appears that an appropriate policy approach might be one that seeks to influence the fundamental forces in the management of critical infrastructure while respecting, wherever possible, the established regulatory and policy framework. With regard to the public information infrastructure, however, it is not clear that existing policy is suited to such an orientation. In Canada, for instance, telecommunications policy for emergency management is rather narrowly focussed and existing programs have been developed based on the predominant response-oriented model of emergency management. This is not to say that existing programs have not proven to be effective within their own fields of endeavour, but only to reinforce my point that a mitigation-oriented approach must adopt a new way of thinking about the management of critical infrastructure if it is to address the more fundamental steps needed to achieve the objective of long-term risk reduction, especially given the complex interdependencies of modern societies.

Certain practitioners in the field of emergency management will argue that mitigation is already taken up in a number of existing public and private programs and areas of research for critical infrastructure. I wish to challenge this claim as it pertains to the public information infrastructure on the basis that many of these programs constitute neither the full extent nor the logical starting point of a thoroughly conceptualized mitigation strategy. In part, the problem lies with the very definition of mitigation, which is hindered by the predominant model currently used in emergency management.

Disaster Process Models and the Problem of Mitigation

A widely accepted model for emergency management is based on a four-stage, recursive cycle comprised of mitigation, preparedness, response, and recovery. Temporally, *mitigation* activities precede *preparedness* activities, both of which happen in advance of an incident and the *response* and *recovery* activities, which follow. Recovery provides the recursive moment in the cycle and a return to mitigation activities. Presumably, a critical incident exists as a future potentiality during the mitigation and preparedness stages, as an actuality during the response stage, and as past history during the recovery stage (see Figure 1.1).

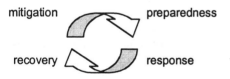

mitigation   preparedness

recovery   response

Figure 1.1 The four phase model of emergency management

Some practitioners substitute the term 'prevention' for mitigation, referring to it as the PPRR model and have traced its origin in the work of the State Governors' Association in the United States in conjunction with a policy initiative known at the time (circa 1978) as 'Comprehensive Emergency Management' (Crondstedt, 2002, p. 10). Since the introduction of this four-stage model, considerable intellectual effort has been made to determine just how these stages are best defined and delineated, with the unfortunate result that very little consensus has so far emerged across disciplines and among practitioners. The term 'mitigation' in particular appears to be used inconsistently by practitioners and researchers alike, sometimes as a synonym for 'prevention' or in other cases conflated with the terms 'preparedness' or 'response.' The problem with the four-stage model is that it leads to conceptual ambiguity that ultimately translates into operational difficulties when determining the specific role and value of 'mitigation' programs. On the one hand, this conceptual ambiguity may hamper efforts to develop consistent standards by which to evaluate and support programs across sectors that might claim to be mitigation-oriented. On the other hand, the inability to define with precision the distinctiveness of mitigation as compared with the other stages of emergency management may result in a failure to shift public policy thinking in any substantially new directions.

Mitigation is a problematic concept in part because it is subject to a vast range of interpretations based on its most common definition as 'pre-meditated action leading to a reduced risk of a loss occurring' or more formally, as in the case of the U.S. Federal Emergency Management Agency (FEMA), 'sustained action taken to reduce or eliminate long-term risk to people and their property from hazards and their effects' (Newton, 2001). The formative terms in most definitions of mitigation are almost always the same:

- pre-meditated
- sustained
- long-term.

Whereas preparedness, response, and recovery represent activities taken just before, during, or after a critical incident has occurred, mitigation seems to imply something just beyond that envelope of actions.[2] Just what that 'beyond' means tends to remain an implicit notion rather than a clearly defined term in the discourse on disaster management. This notion is sometimes expressed as a suggestion that mitigation describes a 'proactive' approach to managing risks, which in my view is essentially the same as saying it is pre-meditated, and does not provide an escape from the critique that preparedness, response, and recovery can also be pre-mediated actions insofar as they are prescribed in planning documents and exercised well in advance of a critical incident. Alternatively, mitigation is sometimes depicted as something that follows a critical incident, providing a linkage between the recovery stage and future preparedness. This is helpful to the extent that it suggests that mitigation provides an important role in societal learning about disasters and human activity.

Despite its apparent simplicity, this model is nonetheless a conceptual trap that presents considerable difficulty when attempting to put the idea of 'mitigation' into practice, raising a number of difficult questions. Is it that mitigation means a 'pre-meditated' effort to improve preparedness or response measures, say through

---

[2] 'Recovery,' however, is unique because it may involve a long-term process of community re-building—sometimes lasting years, as in the case of residents of Kobe, Japan, where recovery efforts stretched to five years after a major earthquake struck the region in 1995 (City of Kobe, 2000).

advanced research to improve weather forecasting or by building new public alerting systems? Or does it mean a 'sustained' program aimed at improved recovery procedures, for example through new forms of insurance and loss coverage? Or does it mean implementing a program of 'long-term' forecasting in terms of risk and vulnerability to hazards in order better target funding for emergency preparedness programs? In operational terms, mitigation is interpreted as a proactive extension of one or more of the other three stages of the model.[3] Perhaps this is an appropriate interpretation of the concept for those directly involved in emergency management. My sense, however, is that the motivation for undertaking formal mitigation programs intimates something that is more ambitious in principle. It seems as if the term mitigation should be an invitation to explore a deeper set of questions related to risk and society, and that such an exploration might eventually result in forms of practical action that are more clearly distinct from the other three phases of critical infrastructure management.

An alternative conceptualization that might invite such exploration begins with an observation by John Newton, a researcher and consultant who has done considerable conceptual work in this area, who suggests that,

> [t]he hallmark of an approach to loss reduction based on a mitigative attitude or approach is the presence of *integrated thinking*—the capability to deal with situations from human, technical and organizational perspectives at various scales of detail. ...

> Part of this attitude requires that we clearly see that natural disasters are not arbitrary occurrences, but rather the interaction of changes in physical systems with existent social conditions. As such, natural disasters can be said to be social phenomena. For without property and casualty losses, concern is much reduced, if we set aside for the moment disruptions to nature, non-human deaths, and the loss of natural resources. (Newton, 2001, p. 2)

If one accepts that the hallmark of mitigation is 'integrated thinking' wrapped around a socially informed conception of disasters, then this would suggest that the predominant model of emergency management may not be the most useful for long-range policymaking. This becomes apparent when one considers common critiques of the wider family of what one consultant has termed 'disaster process' models (Kelly, 1999, p. 25). The wider family of disaster process models are related to the four phase model described above insofar as they share a number of similar features, namely, they tend to be established on four key variables: stage (or phase), specific events, actions to be taken, and time frame. Of these, the temporal dimension is common to all models and its measurement can range from seconds, to hours, to months or years. In many cases, models are based on a linear conception of time and may have difficulty with non-linear or co-constitutive phenomena. Functionalist models are often based on linear time lines with the key

---

[3] I support this claim based on direct observations made while participating in a regional stakeholders meeting for Canada's National Disaster Mitigation Strategy. This meeting was held in Vancouver in 2002 and included participants from all facets of the disaster management community—none of whom it seemed could agree on the meaning of 'mitigation'.

variable of 'actions to be taken' classified according to specific organizational tasks including mobilization, integration, production, and demobilization (Kelly, 1999, p. 25). Events-driven models will often incorporate functional aspects and are based on the use of linear timeframes positioned according to whether something is happening before, during or after a critical incident has occurred. Of course the problem with both functional and events/phase-based models, as critics have noted, is that they establish a linear sequence and then attempt to divide it into clean and often mutually exclusive categories of action. This ignores the various cross-cutting influences and relationships that do not correspond to clear boundaries. Events-based models fall prey to a form of conceptual determinism that 'incorrectly supposes a separation between disaster and non-disaster (i.e., development) periods' (Kelly, 1999, p. 25).

What these models all have in common is that they represent efforts to *disintegrate* the totality of a critical incident through the invention and application of discrete categories using more or less observable phases, events, and actions. In effect, what they do is offer an over-simplified representation of events and conditions that transfers the complexity of a critical incident into the operationalization of the model. In other words, the apparent simplicity of the models betrays the complicated and detailed process of identifying the inputs and measures required to put risk reduction activities into practice. Obscuring the complexity is a chimerical undertaking because it will always reassert itself at the operations level, and because complexity is a function of interdependency. The point at which a disaster occurs is necessarily complex, as it is a result of manifold interactions between social and natural forces. An attempt to remove complexity may be an appropriate method when working within the envelope of actions related directly to emergency management but it is problematic when seeking to look beyond that envelope, as in the case of mitigation

John Newton's call for an integrated and socially informed approach to mitigation seems to suggest that we need to seek out a model that enables a better association between interdependency and the creation and reduction of risk and vulnerability in societies. In other words, a socially informed approach hints at a model that is explanatory rather than simply descriptive. The four-phase model, however, is descriptive inasmuch as it provides a mapping of what we might expect to happen during a critical incident. The incident is assumed as a presence in this model. By contrast, Newton points us toward a way of thinking that does not assume the presence of a critical incident, but rather, to thinking that seeks to explain *how it is* that social and ecological conditions are related to the creation of risk and vulnerability in society. Approaching the question of disasters from this perspective is the first step toward an approach that leads us to investigate the deep social roots of disasters. From a temporal perspective, it expands the time frame of analysis from minutes and hours to years, decades, or possibly even centuries. Likewise, it shifts the unit of analysis away from the narrow confines of emergency management to consider the formation of public policy itself. Newton's call for integrated thinking means adopting a model that grasps the influence of public policy on growth and change in modern societies, a model constructed according to a socially-informed conception of disasters. It is in this conception of disasters as social phenomena where we begin to sense a distinctive meaning for the term mitigation and its relevance in the long-term management of critical infrastructure.

Disasters as Social Phenomena

In contrast to the descriptive function of most disaster process models, an explanatory model seeks to establish a causal relationship based on certain fundamental assumptions. Early in the modern history of emergency management, those assumptions were typically drawn from a military metaphor, or what scholars have labelled the 'patterns of war approach' to natural hazards. The historical context of the United States at the height of the Cold War was deeply influential on disaster-related policy research and planning:

> ... US government institutions provided research funds at that time primarily for studies relevant to understanding the reactions of people to possible air raids. Disasters were viewed as situations likely to elicit reactions of human beings to aggressions and to allow an adequate test of them. ... Bombs fitted easily with the notion of an *external agent*, while people harmed by floods, hurricanes, or earthquakes bore an extraordinary resemblance to victims of air raids. (Gilbert, 1998, p. 12)

This 'patterns of war approach' assumes natural hazards to be external, aggressive forces against which communities have to be fortified in order to defend themselves. The military metaphor, which also played a formative role in disaster management in countries outside the United States,[4] leads to a causal sequence that tends to disregard the social pre-conditions that create vulnerabilities in the face of environmental forces. In terms of suggesting action to mitigate disasters, the approach has been criticized for overemphasizing the natural hazard *qua* aggressor at the expense of examining more deeply rooted social origins of vulnerability and risk (Gilbert, 1998, p. 13). In other words, the patterns of war approach leaves little room for social accountability for disasters, other than to claim that perhaps not enough resources were given over to plan for, or respond to, a *force majeure*. Within this approach, mitigation tends to be interpreted as a form of advanced planning for preparedness, response, and recovery. An explanatory model built on assumptions characteristic of the patterns of war approach leads to a fuzzy conception of 'mitigation' that does not easily stand apart from the other three phases of disaster management. For this reason, one must consider another set of assumptions to inform the explanatory model.

Today the 'patterns of war approach' has been eclipsed, at least in the social sciences, by a more socially aware conception of disasters (Gilbert, 1998, p. 14). According to this alternative view, which emerged from critiques levelled against the military metaphors of a previous era, a turn toward societal structures and

---

[4] For instance, the early history of civil defence in Canada is steeped in the wars (both hot and cold) of the last century. See, for example, Emergency Preparedness Canada (1999). It is no coincidence that an 'aggressor' metaphor has been used in the past to characterize natural hazards, when disaster management has been so closely associated with civil defence.

processes is deemed necessary to questions on the causal aspects of disasters. Researchers adopting this approach acknowledge that *force majeure* is indeed a reality, yet they contend that natural hazards are more appropriately understood as disaster-precipitating events intimately connected with social pre-conditions. Based on this perspective, an explanatory model endeavours to include social pre-conditions rather than to ignore them, thereby drawing attention to the fundamental forces that produce risk and vulnerability in the first place:

> The crucial point about understanding why disasters occur is that it is not only natural events that cause them. They are also the products of the social, political, and economic environment (as distinct from the natural environment) because of the way it structures the lives of different groups of people. There is a danger in treating disasters as something peculiar, as events which deserve their own special focus. By being separated from the social frameworks that influence how hazards affect people, too much emphasis in doing something about disaster is put on the natural hazards themselves, and not nearly enough on the social environment and its processes. (Blaikie *et al.*, 1994, p. 3)

If we include the deep social dimension within our assumptions about disasters then we must also deal with a much longer, and more complicated, causal chain. An explanatory model of disasters based on these assumptions must therefore account for a wide range of influential factors within the social environment and its processes. Along this line of thinking, Blaikie, *et al.* developed a 'Pressure and Release' (PAR) model that reflects this extended causal chain. The Pressure and Release model is based on the view that disasters occur when natural hazards affect vulnerable parts of a community, and that 'vulnerability is rooted in social processes and underlying causes which may ultimately be quite removed from the disaster event itself' (Blaikie *et al.*, 1994, p. 22). The PAR model is useful because it embodies the integrated and socially informed perspective advocated by John Newton and can therefore permit a conceptualization of mitigation that does indeed stand apart from the other three phases of disaster management. Moreover, as I will demonstrate, it helps to guide our thinking when targeting appropriate policy research in the management of critical infrastructure.

The Pressure and Release Model

The causal chain in the Pressure and Release (PAR) model is established by three primary links (see Figure 1.2). Together, these links describe a 'progression of vulnerability' that creates a rising risk of disaster when *root causes* are translated through *dynamic pressures* to create *unsafe conditions*. Within the PAR model a disaster is defined as the product of unsafe conditions confronting the effects of a natural hazard. The 'Release' aspect of the model is meant to emphasize that the root causes and dynamic pressures that lead to risk and vulnerability can also be redirected toward the creation of safe conditions, resiliency, and sustainable community development (Blaikie *et al.*, 1994, p. 219).[5]

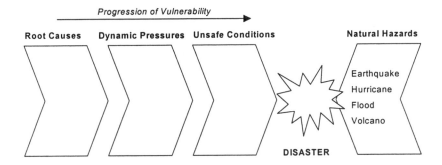

Figure 1.2 Pressure and Release model of disasters

The value of this explanatory model for conceptualizing mitigation is that it encourages integrated thinking because it includes a wide range of influences that are typically quite remote from actual disaster events.   In effect, it extends the sphere of influence backwards in time to focus on distant events that have initiated a momentum or trajectory leading up to unsafe conditions.   For instance, the most distant link in the social chain of the PAR model is the 'root causes' of vulnerability, which Blaikie *et al.* (p. 24) cite as economic, demographic, and political processes reflecting 'the distribution of power in a society' or the prevailing 'ideological order.'   Within the model, the aspect of 'dynamic pressures' represents the next link where the effects of root causes are 'translated into the vulnerability of unsafe conditions.'   Dynamic pressures could be budget allocations, purchasing decisions, resource distribution, or other forms of action that are influenced by root causes.   The last link in the social chain describes 'the specific forms in which the vulnerability of a population is expressed in time and space in conjunction with a hazard.'   Unsafe conditions might include poorly built or maintained infrastructure, unacceptable exposure to natural hazards (e.g., flooding or landslides), insufficient ability to recover losses suffered as a result of a natural hazard, poor health, and lack of awareness and education about natural hazards.

If we are to establish a socially informed concept of mitigation based on the hallmark of integrated thinking linked to pre-meditated action and long-term risk reduction, the Pressure and Release model suggests that mitigation is appropriately situated among root causes and the dynamic pressures that lead to unsafe conditions. I would argue that if a state of 'unsafe conditions' is at issue, then the domain of mitigation more properly yields to that of preparedness, response, or

---

[5] Blaikie *et al.* (p. 9) define vulnerability as 'the characteristics of a person or group in terms of their capacity to anticipate, cope with, resist, and recover from the impact of a natural hazard.'

recovery, where effort is directed at reducing the consequences of such pre-existing conditions.

The PAR model therefore suggests that mitigation is a field of activity properly concerned with the domain of root causes and dynamic pressures. What does this mean for public policy? In this light, the notion of disaster mitigation appears strikingly similar to ideas taken up within the sustainable development movement in which the community planning process itself is opened up to scrutiny as a means of taking into account a wide range of social and environmental factors (Darlington and Simpson, 2001; Geis, 2000; Robinson *et al.*, 1996). Taking this insight one step further, we might then conclude,

> that [if] actual loss reduction occurs through mitigation at [the] community level, *then considering how communities evolve represents an important step toward a fuller appreciation of mitigation*, as an integral contributor to the safer growth of ... society. [emphasis added] (Newton, 2001, p. 6)

A more complete appreciation of disaster mitigation is one that considers it integral to a wider set of constructive social processes, intimately bound to the evolution of communities and directed toward creating safe and sustainable conditions. In this formulation mitigation is clearly set apart from the temporal envelope of disaster process models while maintaining an important continuity with those models. The pressure and release (PAR) model, by contrast, is consonant with Newton's claim that understanding growth and change in communities is integral to mitigation. I wish to press further on this matter, however, by moving beyond an overly simplistic notion of a community as a simple geographical entity in order to include associations or gatherings of similar interests. For instance, those organizations and individuals with a stake in the management of critical infrastructure might be regarded as a community of interest. This distinction is important because the idea that mitigation is intimately linked to growth and change in communities can refer to both physically bound communities, communities of interest (sometimes called 'virtual' communities) and perhaps most importantly, it can refer to the mutual relationships and cross-cutting influences that exist between geographic and virtual communities.

The management of critical infrastructure resides at the intersection between geographic and virtual community interests because it tends to reflect needs arising from local regions while being buffeted by larger societal forces produced by politics, economics, and culture. Mitigation-oriented policy research in this area should thus aim to identify root causes and dynamic pressures, their effects on critical infrastructure, and the extent to which they lead to potentially unsafe conditions. Furthermore, by identifying the deep roots of risk and vulnerability it may be possible to improve management strategies for critical infrastructure through strategic intervention at key moments in time. Figure 1.3 illustrates where mitigation-oriented policy research properly resides within the Pressure and Release model.

The PAR model suggests that an understanding of the fundamental social processes of infrastructure growth and change are necessary if we are to effectively anticipate and avoid the development of unsafe conditions. I might add, by way of clarification, that disaster process models for emergency management remain valid operational guidelines inasmuch as unsafe conditions do and will continue to exist as a result of legacy systems and unanticipated consequences that lie beyond the horizon of foresight. As such, there is room for both types of models in the field, but I am making a case in this book that while mitigation may be a component of the four-phase model, its full conceptualization and implications for public policy are better understood through the different framing that the Pressure and Release model brings to bear on the matter. The PAR model locates mitigation-oriented policy research at the interval between root causes and dynamic pressures in the processes of community development, thereby shifting the focus of analysis beyond the immediate envelope of critical incidents to consider a wider set of social forces within long run scenarios. This fundamental shift in the unit of analysis made possible with the PAR model brings to light that hallmark 'mitigative attitude' advocated by John Newton because it fosters an appreciation of the interdependent qualities of risk and vulnerability in society. Further, it provides an essential starting point for strategizing public interest intervention in the management of critical infrastructure.

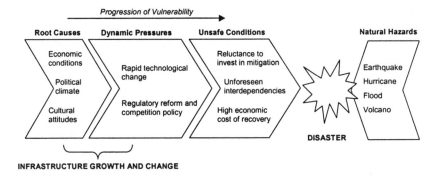

Figure 1.3 Mitigation-oriented policy research

In sum, I have drawn on the PAR model to make a case that the objective of mitigation-oriented policy research as it pertains to the long range management of critical infrastructure should be: (1) to understand the processes by which growth and change occur in critical infrastructure; and then (2) to assess the various means by which to intervene in those processes in order to ensure that long-term risk reduction is given high priority, especially at critical decision points. Yet, the development of an integrated analytical framework to support such an assessment clearly remains challenged by the real limits of current policy frameworks.

The intent of this book is to work toward developing such a framework while also evaluating its feasibility as a policy research methodology more generally. Given this aim, the next matter to address is also conceptual in nature because it requires that I transpose the PAR model into the specific domain of telecommunications *qua* critical infrastructure. As discussed previously in this chapter, the telecommunications sector is a central consideration in the management of critical infrastructure because so many other systems have now become dependent on its continuous functioning. This sector, which is increasingly important for the public information infrastructure more generally, is also undergoing a profound period of transformation with the introduction of competition-based policy frameworks in many countries, the rapid deployment of new technologies, and a growing concern over the threat of terrorist attacks.

Expanding the Conception of Emergency Telecommunications

As one might expect for a field that evolved in close conjunction with national security concerns and disaster planning, 'emergency telecommunications' still embody today many of the features of the so-called patterns of war approach that set the original tone for policy and program design in emergency management. In most countries around the world, much of the focus in the field of emergency telecommunications is confined to addressing needs that arise in conjunction with preparedness, response, and recovery operations. In other words, the field of emergency telecommunications is typically concerned with what I referred to in the previous section as 'the envelope of actions' directly related to a critical incident. By this I mean that the field of practice has been concerned primarily with hardening vital facilities, restoring essential services and maintaining emergency communication channels during an emergency rather than with the long-range management of interdependency and of critical infrastructures in society.

This critique is not by any means intended to diminish the importance of current programs and planning measures. During wide-scale critical incidents telecommunications networks support many important functions ranging from alerting local populations, to coordinating emergency response activities among government and non-government agencies, to enabling the continuity of government functions and business transactions. The GETS (Government Emergency Telecommunications System) program in the United States represents one example of an initiative intended to support this traditional emergency telecommunications role. GETS is an emergency telephone service offered by the National Communications System (NCS), within the Department of Homeland Security, and is designed to provide all levels of government, industry, and non-governmental organizations with emergency access and priority processing in both the local and long distance segments of the Public Switched Telephone Network (PSTN). The service is 'intended to be used in an emergency or crisis situation when the PSTN is congested and the probability of completing a call over normal or other alternate telecommunication means has significantly decreased' (National Communications System, 2004a).

Within the wider field of current practice there is also an important distinction to be made with the term 'public safety' telecommunications, which generally refers to a scenario in which a critical incident does not constitute a community-wide threat *per se* but where an individual seeks to use a telephone to contact emergency services to request assistance on a small scale incident. Today this deceptively simple function is becoming complicated by developments in the telecommunications infrastructure but its essential design remains the same and is based on a single, widely known emergency number—such as '911' or '112'— available in most major urban centres.

While recognizing that there are important and essential distinctions in both policy and practice between 'routine' emergencies and community-wide incidents, and by a taking into account the different kinds of demands each situation might place on a telecommunications network, the generic term 'emergency telecommunications' is sometimes used to refer to the collection of programs, technologies, and services that fall within this broad area of activity. For instance, the European Telecommunications Standards Institute (ETSI) under its EMTEL program takes a unified, comprehensive approach to the field as characterized by its scope of activities:

> The concept of Emergency Telecommunications (EMTEL) addresses a broad spectrum of aspects related to the provisioning of telecommunications services in emergency situations ... [which] may range from a narrow perspective of an individual being in a state of personal emergency ... to a very broad perspective of serious disruptions to the functioning of society ... The concept also covers the telecommunications needs of society's dedicated resources for ensuring public safety; including police forces, fire fighting units, ambulance services and other health and medical services, as well as civil defence services. ... [and to] provide means for dissemination of information to the general public, in particular in hazardous and disaster situations. ... Telecommunications means may also be increasingly used as parts of various community functions such as health services. (European Telecommunications Standards Institute, 2003)

Across this broad range of activities, it is perhaps not surprising to discover that policy research has largely been concerned with issues other than mitigation, as I have characterized it above. In order to transpose the Pressure and Release model into the specific domain of telecommunications it will therefore be necessary to expand the conception of emergency telecommunications to reach beyond the confines of current thinking in the field. The first step toward this expanded conception is to look more closely at the typical range of activities falling under emergency telecommunications and to compare them with the Pressure and Release model.

Using the ETSI classification of emergency telecommunications as a baseline definition, it is possible to identify at least four unique domains of activities involving public institutions:

- Emergency preparedness/national security
- Public safety
- Emergency broadcasting/public alerting
- Public health.

We can expand this list to seven domains if we include three additional categories that involve private and non-governmental institutions:

- International humanitarian assistance
- Lifeline engineering
- Business continuity planning.

In what follows, I will sketch some of the main features of these domains as well as major programs or points of focus within each. As I do so, I will compare each of them with their place in the Pressure and Release model, using this as a basis for assessing their suitability for mitigation-oriented policy research and programs.

The Public Sector and Emergency Telecommunications

Telecommunications for national security/emergency preparedness (NS/EP) is largely concerned with providing continuity of governance following a major emergency or disaster.[6] Its historical roots stem from strategic defence initiatives undertaken during the Cold War and efforts in this area have been renewed in light of perceived threats from terrorist activities. A report published in the early 1990s on the survivability of Canada's public telecommunications networks in the context of a nuclear attack from the Soviet Union is characteristic of this kind of policy research (Hoffman, 1990).

Leadership in this field has come from the United States, where a number of important organizations have conducted research within the context of critical infrastructure protection. Of these, the United States National Communications System (NCS) has initiated numerous projects addressing vulnerability and response readiness for the American federal government (National Communications System, 2004b). In addition, the U.S. National Security Telecommunications Advisory Committee (NSTAC) is a government advisory committee consisting of industry representatives assigned to provide 'industry-based advice and expertise to the President on issues and problems related to implementing national security and emergency preparedness (NS/EP) communications policy.' Similar to the National Communications System, NSTAC produces reports that deal with ongoing developments in the U.S. public telecommunications infrastructure (National Security Telecommunications Advisory Committee, 2003).

After the conclusion of the Cold War, research into national security/emergency preparedness tended to shift its focus toward the impact of new

---

[6] Some readers will note that I have adopted the American term 'national security/emergency preparedness' whereas in other countries such as Canada it is not commonly used. For instance, the Canadian government has created an Emergency Telecommunications bureau within its Ministry of Industry to oversee NS/EP programs. I have chosen to use the American term to make a distinction with the generic notion of emergency telecommunications used by ETSI.

technologies on government preparedness and response capabilities. For example, the U.S. National Institute of Standards and Technology published a study in 1995 on national security/emergency preparedness concerns associated with the Federal Communication Commission (FCC) requirement for 'Open Network Architecture' within the public switched telephone network (National Institute of Standards and Technology, 1995). In Canada, research was conducted to examine the impact of technological and regulatory change on government emergency telecommunications planning and programs (Thomas, 1994). Similarly, governments have funded academic research to look at the various challenges that the Internet presents for emergency preparedness (Anderson and Stephenson, 1997) and wireless communications technology (Anderson and Gow, 2000).

When one examines the body of research dealing within the NS/EP domain it becomes evident that much of it is directed toward objectives of preparedness, response and recovery rather than toward that interval between root causes and dynamic forces suggested in the Pressure and Release model. In part, this focus stems from the current policy framework and approaches used in national security/emergency preparedness settings. In Canada, for example, these responsibilities are set forth in the Federal Policy for Emergencies, the *Emergency Preparedness Act* and detailed at various levels of departmental policy. The lead role for emergency telecommunications responsibilities is assigned to Industry Canada and is described in government policy documents (Emergency Preparedness Canada, 1995). The federal government policy for emergency telecommunications is characterized by a response-oriented approach to support continuity of government, as evident in the responsibilities of Industry Canada, characterized as follows:

- Offering advice and assistance to various levels of government.
- Facilitating provision of telecommunications equipment to government departments as 'required in emergency response operations.'
- Coordinating and managing programs to ensure availability of telecommunications to support emergency responders, to issue warnings, and to meet federal requirements for continuity of governance 'during periods of system overload or degradation.' (Industry Canada, 2002)

The primary role here, as with most NS/EP programs, is to provide a structure for continuity of governance during emergency or disaster situations and as such, research and program development tends to be targeted at that primary function. There is one interesting exception in the Canadian policy that highlights the characteristic fuzziness of 'mitigation' as it is perceived in the field of emergency management. Industry Canada in addition to its preparedness, response and recovery responsibilities is also mandated with,

[p]roviding advice and assistance, as appropriate, to private or public telecommunications undertakings in *mitigating* the disruptive effects of emergencies on domestic and external telecommunications. [emphasis added] (Industry Canada, 2002)

One would think that such a statement should provoke debate about the meaning of mitigation within the context of a national disaster mitigation strategy but no such debate has arisen so far. For instance, Industry Canada considers a number of its undertakings as mitigation programs because they are seen to serve a role in supporting effective response activities during a critical incident, thereby reducing potential losses in the face of a disaster.[7] I would suggest that such claims simply underscore the point I made earlier about the conceptual ambiguity attached to the notion of mitigation as it comes to be interpreted by various stakeholder groups. It is not that I wish to critique such claims except to point out that they represent a very narrow interpretation of the term that does not correspond to the wider considerations raised by the Pressure and Release model.

The Private Sector and Emergency Telecommunications

Emergency telecommunications support for the private sector has tended to exist as an issue for the marketplace and thus has remained outside the domain of public policy. Its private market equivalent is known as 'business continuity planning' or 'business resumption planning.' The literature on business continuity planning has taken up the challenge of telecommunications failures, most notably reflected in the work of Leo Wrobel, who has written extensively in this area (Wrobel, 1993, 1997). Wrobel's work addresses the needs and issues confronting business and community organizations in the face of telecommunications disasters, and he has produced detailed planning guidelines to support business continuity planning. The thrust of Wrobel's message is that private organizations must prepare themselves by being well informed and by establishing good working relationships with their telecommunications carriers. In fact, the field of business continuity planning is important insofar as it does demonstrate a concern with the idea of mitigation, which is expressed through its emphasis on advanced planning for continuity of operations. Continuity refers not only to the capability of a private firm or community organization to quickly respond to a disaster but, more importantly, draws attention to the idea that long- range planning for successful recovery is a means of mitigating disasters. In reviewing the literature in this field, however, one discovers that focus tends to be on existing solutions and methods rather than on the root causes and dynamic pressures affecting an organization's ability to effectively engage in continuity planning.

Despite the economic threat posed by the growing interdependency of critical infrastructure in modern societies, there remain only a few published accounts that describe and quantify the vulnerability of the business sector in the event of a mass emergency or disaster. Perhaps most significant within this small body of research is a series of surveys undertaken by the Disaster Research Center (DRC) located at

---

[7] The programs I include here are Priority Access to Dialing (PAD) and High Probability of Completion (HPC). Together, they are similar to the American GETS program. It was an Industry Canada representative who expressed to me the opinion that they are 'mitigation' programs inasmuch as they had been accepted as such by an inter-governmental committee conducting preliminary work on Canada's National Disaster Mitigation Strategy.

the University of Delaware (Nigg and Tierney, 1995; Webb, Tierney and Dahlhamer, 2000). While these studies do not address the impact of communications networks specifically, the research does concern itself with disruptions in critical infrastructure facilities, and findings generally indicate that a significant portion of losses and business closures result not from the disaster-precipitating event itself but from a range of secondary impacts due to failures in telecommunications, electricity and other supporting infrastructure.

In contrast to the DRC studies, there are ongoing civil engineering studies on infrastructure vulnerability that examine the direct physical impact of disasters on communications networks, yet only touching upon the wider socio-economic fallout (Schiff and Tang, 1995). This body of research into what is termed 'lifeline engineering' represents another form of private sector/non-governmental involvement in emergency telecommunications. While this area of investigation may have a longer range focus than the critical infrastructure protection field, it deals primarily with physical network elements rather than wider social and economic concerns. Upon review of this body of research, I found only one specific study aimed at measuring the wider social costs of a telecommunications network outage, which was undertaken as a case study following a cable fire in Setagaya, Japan, in the late 1980s (Takansashi *et al.*, 1988). Details from the Setagaya case are not relevant to this study aside from the fact that it was an early, 'pre-Internet' case that merely hints at the potential impact of a widespread failure in today's highly integrated public information infrastructure.

The Management of Critical Infrastructure

Despite the commendable work done in the field of emergency telecommunications in both the public and private sectors, there remains a significant oversight; namely, that most *policy research* has been confined to the national security/emergency preparedness domain. Recently there has been research done to examine the regulatory dimensions of disaster preparedness for telecommunications in developing nations (Samarajiva, 2001; Srivastava and Samarajiva, 2001) but while representing an important starting point, it is only preliminary work and much more needs to be done in the area. Work has also been done in the domain of international humanitarian assistance in conjunction with the UN's Working Group on Emergency Telecommunications and its efforts behind the Tampere Convention (Harbi, 2001). If one draws comparisons across the domains of emergency telecommunications, however, it becomes evident that research is predominantly focussed on dealing with unsafe conditions through improved risk management, enhanced preparedness, or the development of advanced planning and recovery strategies. For instance, the domains of lifeline engineering and business continuity planning adopt a somewhat more proactive approach to emergency planning (or some would argue 'mitigation' as per my previous points about its ambiguity) but research in neither domain deals specifically with the fundamental processes that influence the shape of the telecommunications infrastructure in the first place. It is also evident among public sector interests in the national security/emergency preparedness domain that the predominant focus of research is on how best to harden current systems against attack, much in the vein of the 'patterns of war approach.'[8] Within both public and private sectors, policy research seems to have largely ignored the root causes and dynamic pressures that influence growth and change in critical infrastructure.

A comparative analysis of the typical domains in emergency telecommunications with the PAR model seems to indicate that current research and programs fall within the envelope of actions most closely associated with emergency management, yet consideration is not given to the possibility that mitigation might exist outside this envelope. This is a significant oversight, given the rapid deployment of new technologies within the wider context of global telecom reform and information society initiatives. Among other things, telecom reform has introduced competition into former monopoly markets, leading to new dynamic pressures in the form of radical restructuring of the sector, the appearance of new stakeholder groups and investors, and a dramatic increase in spending and attention given to innovation (Bauer, 2003). On the one hand, these dynamic pressures produce new opportunities that include flexible and cost-effective telecommunications services to support emergency management.[9] On the other hand, they also produce hidden risks and vulnerabilities when rapid technological developments, impulsive investment, and uncoordinated decision-making processes lead to the formation of critical path dependencies in network infrastructure. Making the most of innovation and reducing the long-term risks associated with a rapidly changing infrastructure means investigating the deeper relationships behind these processes, including economic conditions, political climate, and cultural attitudes.

Having now been transposed into the Pressure and Release model, the concept of emergency telecommunications can be expanded to include a study of the wider social forces that confront dynamic pressures in the telecom and communications sectors to produce unsafe conditions. Unsafe conditions include reluctance to invest in mitigation, unforeseen infrastructure interdependencies, and potentially high human and economic costs of recovery following a critical incident. However, within this expanded conception, the term 'emergency telecommunications' sheds its intuitive value because our concern is no longer with emergencies *per se* but rather with the assessment and long-term management of large technical systems. I would propose therefore that it is at this point, located at the intersection between unsafe conditions and their social production, that we cross a logical and semantic boundary from the domain of 'emergency telecommunications' into the wider field more suitably termed 'management of critical infrastructure.'

---

[8] This perspective also extends to research in the emerging domain of critical infrastructure protection, particularly that dealing with cyberterrorism (see Cordesman, 2002).
[9] The European Telecommunications Standardization Institute (ETSI) EMTEL website provides some examples of work taking place to use new technologies to support emergency communications. http://www.emtel.etsi.org/overview.htm

Summary

This chapter began with the observation that mitigation has emerged to become an increasingly important aspect of emergency preparedness policy in countries around the world. As a result, one might expect that it will also come to play a more prominent role in public infrastructure development, including telecommunications networks and services. In most cases, however, very little discussion of substance has been given over to critical infrastructure within the scope of mitigation strategies.

As a first step in that direction, I proposed an explanatory model of disasters in order to expand the conception of mitigation to include everyday processes of community development—processes often far removed from what many would consider 'emergency management' but which are nevertheless deeply implicated in the production of risk and vulnerability in society. In other words, mitigation for purposes herein is taken as inherent to fundamental forces of social and technological change, rather than as events more directly associated with critical incidents.

The chapter concluded with a critique of traditional approaches to 'emergency telecommunications,' where I argued that they do not address 'mitigation' as it might be interpreted using the Pressure and Release model. This explanatory model of disasters places emphasis on understanding root causes and dynamic pressures as driving forces behind a progression of vulnerability in society. My intent here is to argue for an expanded conception of the field into a realm more properly termed management of critical infrastructure. Such transformation in policy thinking requires a substantial focus on the forces of growth and change influencing the development of critical infrastructure.

Based on this approach, the policy research in this area should set as its objectives: (1) to understand the processes of growth and change in critical infrastructure development; and (2) to assess the various means by which to intervene in those processes in order to ensure reconciliation of public interest with the need for a sustained program of risk and vulnerability reduction. The next chapter introduces an analytical framework intended to contribute to the fulfilment of these objectives.

Chapter 2

The Design Nexus

Design Thinking

The emergence of mitigation-oriented public policy indicates a growing concern with the root causes of risk and vulnerability in society, a concern informed by an understanding that community sustainability is assured through careful coordination of social and technological change. Integrated thinking, as I have argued, is the hallmark of coordinated action, reflecting an acute awareness of the interdependency dilemma that modern societies face as they become increasingly interconnected through the public information infrastructure. Mitigation strategies are therefore appropriately directed at elementary processes of community development, which means in turn that the research efforts should focus on understanding how growth and change occur in these systems. The Pressure and Release model helps us to see that mitigation-oriented policy research is positioned at the interface between root causes and dynamics forces, where risk and vulnerability can be 'headed off at the pass,' as it were, to reduce the incidence of unsafe conditions in the first place.

A basic assertion of mitigation-oriented thinking is that it is possible to be safe by choice and that we are well advised not to leave such things to chance. Underlying this assertion, of course, is the assumption that choices are possible. While it is true that choices are not always possible, the importance of choice as the basis for public policy decisions need not be diminished: 'There is no area of contemporary life where design—the plan, the project, or working hypothesis which constitutes the "intention" in intentional operations—is not a significant factor in shaping human experience' (Buchanan, 1996, p. 6). Mitigation-oriented thinking highlights the importance of the decision-making process in the management of critical infrastructure and attempts to expand the domain of intentionality germane to these undertakings. The mitigation strategy is a form of practical action toward this end and thus embodies a form of *design thinking*, which according to design theorist Richard Buchanan, is marked by 'a concern to connect and integrate useful knowledge from the arts and sciences alike, but in ways that are suited to the problems and purposes of the present' (Buchanan, 1996, p. 4). Buchanan's use of the term 'design thinking' is intended to emphasize both the practical and ethical implications of human intentionality in every corner of contemporary life:

> Design ... has a *technologia*, and it is manifested in the plan for every new product. The plan is an argument, reflecting the deliberations of designers and their efforts to

integrate knowledge in new ways, suited to specific circumstances and needs. In this sense, design is emerging as a new discipline of practical reasoning and argumentation ...(Buchanan, 1996, p. 18)

Design thinking is a primary form of integrated thinking and provides an operational approach to mitigation-oriented policy. This is especially the case if we remove it from a narrow definition of industrial arts or aesthetic ornamentation to embrace it as the force of human intentionality behind all enduring works of art and science. Arguing for the importance of linking design and public policy, Langdon Winner suggests that 'special care must be taken in the fashioning of all things built to last' (Winner, 1995, p. 150).

Put another way, critical infrastructures are *designed* through the interactions of a large number of stakeholders in order to function as interconnected, interoperable networks of technologies and services. Not only are these large technical systems 'built to last' but they may also span vast geographical reaches. The public switched telephone network (PSTN), for instance, is considered the largest integrated technical system ever created (Karlsson and Sturesson, 1995). In contrast to standalone artifacts, however, critical infrastructures and their associated networks are usually highly diffuse systems, which makes it difficult to say at the outset where or by whom they are 'designed,' raising the related problem of establishing analytic boundaries around such complex and interdependent systems.

I suggest therefore that it may be helpful to think of the evolution of critical infrastructures as the product of a *design nexus,* involving coordinated interactions between quasi-independent organizations, international bodies, national regulatory agencies, national standards bodies, service providers, equipment manufacturers, and any number of other interested parties including (potentially) a diverse community of users. While each of these organizations is principally concerned with its own mandate and agenda, in the broader context it is their collective action that supports the ongoing development, deployment, and maintenance of critical infrastructure within individual countries and throughout the world. In this chapter I will explore various aspects of this distributed design process by adopting an approach to technology assessment specifically concerned with the design nexus.

Technology Assessment and the Design Nexus

Technology assessment is a research strategy prompted by a desire to understand and to anticipate the interaction of technology and social forces in order to inform policy making in both public and private institutions (Herdman and Jensen, 1997; Hill, 1997; La Porte, 1997). As an approach to policy research, technology assessment is probably best understood as a flexible arrangement of theory and method that emphasizes one of three aspects: a particular technology project, a technology-related problem (i.e., negative externality), or the social impact of innovation (Porter, 1980, p. 51). I describe technology assessment as a 'flexible arrangement' because specific technology assessments will take many operational forms depending on the requirements and constraints of a given study (Armstrong and Harman, 1980; Nguyen *et al.*, 1996; Porter, 1980). I have also chosen the term to deliberately convey the rich qualitative character of technology assessments where 'the researcher is faced with "fuzzy" issues underneath which lie multiple, often broad research questions, and he or she is asked to address them in a rather "messy" or uncontrolled environment' (Hedrick, Bickman and Rog, 1993, p. 3). Technology assessment is conducted not in the laboratory but in the wilderness of the world, and as such, adaptive and iterative processes are required to produce meaningful results.

In the mid-1990s, a number of books and articles were issued to address various methodological aspects of technology assessment. Findings tended to contrast the particular historical circumstances of the development and evolution of technology assessment in North America with that of Europe, where a growing emphasis on technology design had begun to flourish (Bereano, 1997; Berloznik and van Langenhove, 1998; Van Den Ende *et al.*, 1998; Wood, 1997). While forward thinking proponents of technology assessment in Europe appeared to be the first to view *design* as the critical nexus between technology development and social policy, the idea had also taken root within North American technology studies. Influential American academics were also theorizing design as the active process of embedding social norms and political economic motives into technological systems, a process which Langdon Winner labelled 'political ergonomics' (Winner, 1995). Philosopher Andrew Feenberg also pointed to design as the foundation for a new democratic politics of technology (Feenberg, 1999) and Mansell's political economic research drew attention to the design of telecommunications networks as a pivotal enabling and constraining factor in both European and American settings (Mansell, 1990, 1993, 1996, 1999). Historical research on the development of large technical systems also suggests that materiality and design are critical variables in shaping the influence of these systems on the social world through a form of 'bias' or 'momentum' (Hughes, 1983, 1987; Innis, 1951, 1986).

Constructive Technology Assessment

Drawing public policy into the design nexus is a matter for Constructive Technology Assessment (CTA), an approach spawned by research on technology dynamics done by the Dutch Science Dynamics program. CTA presents an important contrast with the so-called 'early warning' approach of technology assessment as it developed in North America during the 1960s and 1970s. Whereas proponents of early warning technology assessment such as the United States Office of Technology Assessment (OTA) tended to adopt an *exogenous* model of technology development for its research projects, the CTA program was conceived within a participatory *endogenous* model of technology development (Eijndhoven, 1997). An exogenous model anticipates a finished technology that enters into and creates effects in a society (Edge, 1995). Endogenous models, by contrast, recognize that a technology and its effects are not dropped 'stork-like' into a society but are produced, womb-like, by various interested parties through deliberation. An endogenous model of technology development encourages CTA practitioners to view technology projects as highly contingent undertakings open to a range of alternative possibilities, especially in the early stages of development.

A Dutch government policy memorandum of 1984 is credited with formulating the overall goal for CTA, although it never referred to CTA as such. 'The Memorandum,' write Schot and Rip, 'argued that the function of TA studies should be to let societal aspects become additional *design criteria*' [emphasis added] (Schot and Rip, 1996, p. 252). In the tradition of constructivist approaches, CTA redefines technology assessment as an active contribution to the process of design as opposed to an independent program of technology impact analysis. The Dutch policy memorandum also established the Netherlands Organization of Technology Assessment (NOTA), which later became the Rathenau Institute. NOTA developed the new TA approach along two major paths: first, by introducing public participation in the TA process; and, second, by funding several key studies looking at the value of TA at the design stage of technology development (Schot and Rip, 1996, p. 253).

Since the Memorandum was issued, the Constructive Technology Assessment approach has been recognized by numerous organizations across Europe including TA groups in Denmark, Germany, Norway, and the Organization for Economic Cooperation and Development (Schot and Rip, 1996, p. 254). In addition, various EU programs in research and development have recognized the ideas within CTA as contributions to improved technology policy and development (Berloznik and van Langenhove, 1998). It is therefore reasonable to claim that CTA has become a respected approach in the technology assessment field, particularly in the European community. In North America its presence is less evident—likely a result of the U.S. Office of Technology Assessment (OTA) dominating the field until it was disbanded in the mid 1990s. The philosophy behind CTA, however, has begun to circulate among researchers and policymakers in both Canada and the United States, gaining prominence in a number of selected areas of science and technology policy research (Loka Institute, 2004).

In its most general formulation, CTA is taken to be an input to technology design with an emphasis on iterative and contingent processes or 'societal learning' as a major outcome (Eijndhoven, 1997, p. 280). In effect, the endogenous model of technology development and process-oriented principles establish twin foundations for CTA practitioners (see Figure 2.1) and make an innovative contribution to technology policy research:

> CTA can be seen as a new design practice ... in which impacts are anticipated, users and other impacted communities are involved from the start and in an interactive way, and which contains an element of societal learning. (Schot and Rip, 1996, p. 255)

Schot and Rip summarize the strategic view of CTA practitioners when they claim it is an approach that 'attempt[s] to improve our chances to arrive at better path dependencies by broadening technological design and development' (Schot and Rip, 1996, p. 258). In other words, Constructive Technology Assessment practitioners seek to intervene in the design nexus to explore the potential impacts of technology projects well before they become entrenched or 'locked-in' through investment and uptake.

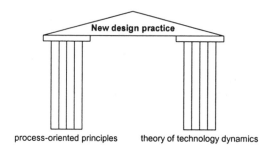

process-oriented principles   theory of technology dynamics

Figure 2.1 Twin pillars of CTA as a form of design practice

The Collingridge Dilemma

For all its successful projects, the experience of the U.S. Office of Technology Assessment revealed serious limitations to the 'early-warning' approach to technology assessment. In particular, it soon became evident that future *effects of* and *influences on* technology are not easy to determine and pose an inherent challenge to any approach that seeks to grapple with the early design phase of a technology project.[1] Efforts to anticipate future outcomes of a technology project must address what has come to be known as the Collingridge Dilemma, which is described as follows:

> The Collingridge Dilemma is often used to refer to the fact that forecasting future effects of technology is difficult, whereas well-developed technology is difficult to direct, because it has become embedded in society ... The Collingridge Dilemma points to the fact that the early warning function of technology assessment has severe limitations, because either the knowledge or the power are missing to change the direction of technological development, leaving quick adaptation to new technology as the only way society can react. (Eijndhoven, 1997, p. 279)

Another view on the matter describes the Collingridge Dilemma in more succinct terms:

---

[1] This challenge speaks to a larger theoretical problem integral to technology studies. For instance, if one rejects technological determinism outright when dealing with the problem of anticipating future effects, then one is obliged to assume either the totality of the social as an influence, or the hybrid possibility that technology and society are somehow co-constitutive. The former amounts to a radical position of social determinism that critics have argued is untenable (Sismondo, 1996).

First, [there is] an information problem: impacts cannot be easily predicted until the technology is extensively developed and widely used. Second, a power problem: control or change is difficult when the technology has become entrenched. (Rip, Misa and Schot, 1995, p. 7)

The Collingridge Dilemma reflects the double bind of path dependency: effects are difficult to anticipate until a technology is deployed, yet once deployed, it may be impossible to affect substantial change to it because of investments made in its development and deployment (see, for instance, David, 1986). David Collingridge himself suggested that a solution to this dilemma could be found in developing highly flexible technology designs capable of multiple configurations. Collingridge advocated an approach to building technology that is easily adaptable to a variety of social settings and different user needs. CTA proponents argue that this is a conceptually flawed strategy because some degree of 'entrenchment is necessary to implement a technology' (Rip, Misa and Schot, 1995, p. 7). In other words, the notion of full flexibility is problematic in practice because it overlooks the inherently value-laden and goal-oriented processes within each progressive phase of a technology project. Researchers have observed, for instance, that flexibility *must be constrained* at numerous points in the design process in order *to effect closure* on any technology. This is, after all, the very point at which the entrenchment or embedding of values and path dependency begins. Without minimal degrees of closure along the way there can be no movement whatsoever toward a functional technological artifact. CTA proponents accept this logical necessity as integral to technology design but assert that policymakers must first attempt to understand the implications of a proposed design and then participate in exploring alternative pathways by generating a range of alternatives:

The effect of CTA [is] not to bring technology under control so that it plays a less dominant role in society. What changes is the form of control and how technology development is played out. ... The goal is to anticipate earlier and more frequently, to set up design processes to stimulate reflexivity and learning, and thus to create greater space for experimentation. (Schot, 1998)

In order to meet the methodological challenge presented by the Collingridge Dilemma, CTA practitioners must first adopt a perspective on technology dynamics that enables them to identify the moments, means, and motivations by which closure and entrenchment happen amidst the otherwise complex ambiguities of technology design. This is an *analytic undertaking*, which has been studied under the labels of Social Construction of Technology (SCOT) and Actor Network Theory (ANT). In conjunction with the analytic project, however, there is also a *political undertaking* that needs to be addressed within CTA in order to address questions of stakeholder composition, involvement, and intervention to ensure that all 'appropriate' perspectives and concerns are taken into consideration. This leads logically to a *normative undertaking*, which is necessary to address questions of social values and cultural practice in deciding the criteria of 'appropriateness' for society. These three undertakings are a response to the methodological challenge posed by the Collingridge Dilemma, and each is integral to the potential achievements and limitations of CTA (see Figure 2.2).

Figure 2.2 Three undertakings of CTA

Analytic Undertaking: Generalized Principle of Symmetry

One prominent researcher in the field has written that 'from the beginning, CTA emphasized the necessity of insight into, and analysis of, the dynamics of technological development and how technology gets embedded in society' (Rip, 1994). It is this emphasis that has drawn CTA practitioners to constructivist science and technology studies for a theoretical perspective on the endogenous model of technology dynamics. Among other things, CTA takes from science and technology studies the generalized principle of symmetry as a foundation for its model of technology dynamics (Rip, 1994). In its original form, established by a research program under the label Social Construction of Technology (Pinch and Bijker, 1987), the principle of symmetry means that all relevant social influences on a technology project are to receive equal analytical weighting—not just those influences that are perceived to be responsible for the successful achievement of a technology project, as had been commonplace in earlier histories of technology (Cutcliffe, 2000). Actor Network Theory developed a more radical position, arguing for a *generalized* principle of symmetry, where it is claimed that the very categories of technical and social should not be assumed *a priori* but should be considered as part of a process of classification or 'purification' necessary for 'successful' technology projects (Callon and Latour, 1981; Latour, 1993). Callon, for instance, has argued that 'the distinction between the technical and the social is the result, not the cause, of the stabilization of socio-technical ensembles' (see Bijker, 1995). For this reason the approach of actor network theory is called 'relational materiality' by Law, who has described it as 'a ruthless application of semiotics … [that claims] entities take their form and acquire their attributes as a result of their relations with other entities' (Law, 1999, p. 3). In this formulation, the distinction between 'technical' and 'social' is taken to be a relational effect that is uncertain, reversible, and derived from the contingent relations and forces found in a given technology project.

In operational terms the generalized principle of symmetry suggests that analysis should focus on the most nascent stage of a technology project—the initial conception phase, where the act of constructing relational entities can be observed in action. To some extent this method circumvents the Collingridge Dilemma because it does not treat the technological and social as mutually exclusive but recognizes that design problems deal with 'quasi-objects' consisting of heterogeneous and often indeterminate elements, out of which specific artifacts come to be stabilized (Latour, 1993, p. 51). Perhaps not coincidentally, scholars in the field of design studies have expressed similar ideas in this regard:

> In actual practice, the designer begins with what should be called a *quasi-subject matter*, tenuously existing within the problems and issues of specific circumstances. Out of the specific possibilities of a concrete situation, the designer must conceive a design that will lead to *this* or *that* particular product. A *quasi-subject matter* is not an undetermined subject waiting to be made determinate. It is an indeterminate subject waiting to be made specific and concrete. (Buchanan, 1996, p. 16)

Buchanan refers to this as the 'wicked problem' of design. The wickedness derives from the fact that designers are not discovering but *inventing*, and often within an indeterminate field of possibilities. Design is therefore an integrative, constructive discipline that draws from a wide range of sources including 'signs, things, action, and thoughts' to organize experience and develop design solutions (Buchanan, 1996, p. 8). Put another way, design is an act of 'heterogeneous engineering' (Callon and Law, 1997) in which both animate and inanimate actors come to play an active role in the creation of a new technological achievement. Latour's term 'quasi-object' and Buchanan's term 'quasi-subject matter' both describe that initial moment of indeterminacy that precedes the translation of interests into specific forms of technological artifacts and practices. Latour describes this indeterminacy in literary and philosophical terms with his claim that '[b]y definition, a technological project is a fiction, since at the outset it does not exist, and there is no way it can exist yet because it is in the project phase,' adding that 'no one is a Platonist where technology is concerned' (Latour, 1996, p. 23).

Drawing on the generalized principle of symmetry, CTA practitioners therefore consider the design nexus to be a negotiating space wherein future effects are sought after by means of an intentional process of heterogeneous engineering, which is an observable discursive process between interested parties. I will expand on this idea further in the chapter, but it is important to acknowledge that this notion of 'sought after' effects also introduces a related question of *unintended* effects. In many respects it is precisely the question of unintended effects that is of interest to CTA practitioners, as it presents a major obstacle to traditional forms of technology assessment and is a central concern in political debates about new technologies.

Political Undertaking: Participatory TA

Constructive Technology Assessment is an offspring of a related field known as 'parliamentary technology assessment' in which technology development is viewed as political activity worthy of public reflection and debate (Eijndhoven, 1997, p. 270). Drawing on these origins, CTA contributes to technology policy a strong emphasis on rational policymaking through public participation and debate, leading to what is termed in the literature, 'symmetrical social learning' (Schot, 2003, p.274), and in this emphasis it is possible to identify close affinities with the political theory of deliberative democracy (Benhabib, 1996).   CTA invites participation among many stakeholders with the optimistic view that such participation will expand both the breadth and depth of debate about new technologies. With this hopeful objective in mind, however, CTA practitioners are faced with practical problems on two matters. First, who are the stakeholders to be represented? Second, how is debate to be framed or contextualized?

The matter of stakeholder representation is one of deciding on the constituency that is to be invited to participate in technology assessment and the problem stems in part from the meaning of the term 'participation' as it pertains to a given setting. For some CTA practitioners the idea of participation means public or 'citizen' participation that 'encompasses a group of procedures designed to consult, involve, and inform the public to allow those affected by a decision to have an impact into that decision' (Rowe and Frewer, 2000, p. 6). There are two operative terms in this statement: 'those affected' and 'have an impact'. Recalling a central problem of the Collingridge Dilemma, how are we to determine those who are likely to be affected if we take into account the inherent ambiguity of a technology project in its conception phase? Moreover, what does it mean to talk of having an impact on proceedings when, as Rowe and Frewer (2000, p. 6) have noted, participation can range from one-way public information campaigns, to extensive dialogue, to forceful intervention?

In practice, the ideal CTA situation seems to be one that is characteristic of the consensus conference, a European invention whose development is co-extensive with the birth of CTA. The consensus conference 'is a structured education, discussion, and decision-making process designed to inform citizens about a subject in question, and then to elicit their informed judgments about it' (La Porte, 1997, p. 208).   Richard Sclove, an American advocate for the consensus conference, describes it as an extended commitment of time that includes selected members of the public in extensive preparation and intimate involvement in the assessment process (Sclove, 1996, 1999).

The consensus conference may provide participants with some influence over the process of technology development but the basic question of stakeholder representation remains and will need to be addressed in each unique set of circumstances. Again, this is complicated by the double bind of the Collingridge Dilemma. In theory, CTA practitioners address this challenge through a process-oriented approach, whereby successive assessments are adjusted to new conditions as learning progresses. As such, the problem of 'those affected' can be dealt with through repeated sessions where new conditions and higher or more specific levels of learning in an assessment project lead to an evolving mix of stakeholders. This evolving mix could include previously overlooked stakeholders or it could shed those stakeholders with decreasing relevance to the project. Of course, this would require some form of ongoing commitment to symmetrical learning among stakeholders and organizers alike.

Framing the debate, on the other hand, involves the problem of stakeholder congruency, which refers to the establishment of minimum conditions of shared understanding among participants who may otherwise have divergent expertise, backgrounds, and interests (Grin and Graaf, 1996, p. 77). Congruency is also the vital foundation for the learning aims of CTA. How can congruent terms of reference be reached as an integral step in the assessment process? Following the prescriptions of CTA, a number of approaches have been developed to address the problem of stakeholder congruency, including a technique that involves the study of metaphor in the design process (Mambrey and Tepper, 2000). These approaches are generally directed at a three-step process for achieving congruent frames of meaning among stakeholders: articulation, reflection, and a limited degree of consensus building. In the field of science and technology studies, this problem of stakeholder congruency is related to Bijker's concept of the 'technological frame' (introduced in chapter six). It is important to note that stakeholder *congruency* is not synonymous with *consensus*, despite the fact that the latter may be the desired objective of a specific technology assessment. I make this distinction to highlight a difference between an orientation toward a common problem or concern (congruency) versus an agreement on the terms and conditions of further action (consensus). In some cases, the aim of a technology assessment may not be to establish consensus but rather to provoke further debate. In each case, however, there remains a basic need to contextualize the setting in which participation is to take place, whether or not the stakeholders agree on the most appropriate course of action. The challenge of stakeholder congruency may be partly one of language usage and expertise (as with difficulties that might arise between engineers and lay persons) but in certain circumstances the barriers might extend into the domain of ethics and values, or what is sometimes called the field of 'normative' perspectives.

Normative Undertaking: Directed Incrementalism

While CTA's kindred field of research in science and technology studies has been criticized for failing to take a normative stance on technology development (Winner, 1993), CTA practitioners have adopted at least an implicit normative position on the matter, 'embrac[ing] values such as being anticipatory, reflexive, and oriented toward learning' (Schot and Rip, 1996, p. 265). Some critics have argued that this kind of normative stance borders on an empty value system (Radder, 1996) but its proponents reply that it should be recognized for embracing open-ended democratic debate that avoids imposing prescribed values on stakeholders and their positions.

To counter charges of relativism in such a meta-level value system, Grunwald (2000) has proposed that the principle of symmetrical social learning could be reconciled with wider social values by shifting the criteria of acceptance in technology assessments from substantial measures to procedural measures. In effect, this would mean working toward consensus not on the content of particular technologies *per se*, but rather on the *criteria of acceptability* for technology development. In the case of CTA, these procedural criteria could be applied to a wider set of societal considerations including the question (as taken up in this book) of making long-term risk reduction an input to critical infrastructure development. For Grunwald, criteria of acceptability are established on the basis of pragmatic rationality, which is a theoretical concept beyond the scope of our current investigation. A cursory understanding of the concept, however, does indicate an attempt to establish criteria of acceptability based on what Grunwald (2000, p. 120) terms 'the social contract' or 'the medium of society [that enables] continuity, stability, and identity beyond the individual level,' which he translates into an operational model based on 'directed incrementalism.'[2] The strategy of directed incrementalism yields some insight on the normative dimensions of CTA and the associated challenges posed by the Collingridge Dilemma.

Contrasting long-range strategic planning with short-range tactical planning, Grunwald takes both to be unacceptable models for reconciling flexibility with long-term objectives. Detailed technocratic models, for instance, posit a stable future objective and establish a pre-defined trajectory to reach that objective. In effect, they presume 'the plannability [*sic*] of large areas of society at the macro-level' and ignore the possibility that societal conditions themselves may change and, as a result, that the original objective may become obsolete or problematic in the future (Grunwald, 2000, p. 128). On the other hand, Grunwald notes that incrementalist models posit few long-term objectives, 'staggering through history like a drunk putting one disjointed incremental foot after another' (2000, p. 109). In the incremental model, long-term objectives such as those associated with sustainable development and disaster mitigation are difficult to achieve because there is no effective means of establishing distant targets on which to guide long-range technology development. 'Directed incrementalism' is considered the middle road through this dilemma. Using this approach, Grunwald suggests long-range targets could be established as a second-order set of criteria. This is allows for incremental changes in specific targets but draws on these changes to feed back into the assessment and revision of long-term objectives:

> Permanent reflection on the goals and the means to attain them leads to incremental changes of direction in the development, of the goals as well as of the measures to reach the goals. The change, however, does not show an erratic behaviour. This kind of development allows [us] to get closer to the envisaged area of goals and to take into account the short-ranged flexibility requirements (which are leading to the incremental changes of direction). (Grunwald, 2000, p. 129)

With a strategy of directed incrementalism, long-range targets are taken to be second-order criteria, meaning that the short-range feedback process is used on the one hand to help the system regulate itself and on the other hand to provide input for controllers to reflect upon and learn from the current state of the system, to formulate new long-term objectives for its development, and to make changes

---

[2] Much work in CTA and Grunwald's in particular has affinities with the ideas of Habermas and Ulrich Beck, both of whom have dealt with the question of rethinking modernity in the face of the postmodern challenge. Beck in particular has examined the concept of risk and modernity.

directed toward those new objectives.[3] This notion of second-order reflection appears to be at least an implicit idea behind symmetrical social learning as it is conceived within CTA. Grunwald's model is helpful for understanding the so-called *broad* and *deep* learning that CTA proponents refer to in the literature, which is meant to minimize unintended consequences in the deployment of new technologies (Schot, 1998). Broad learning describes a wide, sweeping exploration of possible links between design, user demands, and issues of social and political acceptability. This is typified by a group of stakeholders sharing ideas and increasing awareness of each other's positions on a matter. Deep learning, on the other hand, is transformative with a first and second order component. First order deep learning improves the ability to deal with basic tasks at hand (i.e., procedure). Second-order deep learning is related to the more complicated requirements of stakeholder congruency and normative perspectives because it 'requires clarifying values and ways of relating values to each other' (Schot and Rip, 1996, p. 257). When successful, second-order deep learning may lead to wholesale adjustments in held ideas, values, and practices.

CTA and the Principles of Mitigation

A comparison between the principles and objective of Constructive Technology Assessment and those embodied in a mitigation-oriented policy framework such as Canada's national mitigation strategy reveals that both approaches share a number of important features. In the analytic domain, the mitigation strategy and CTA share a concern for the basic processes of design in social and technical systems and both endeavour to bring an endogenous perspective to systems undergoing growth and change. Likewise, mitigation strategy and CTA both seem to share a respect for participatory politics that seek to cultivate wide stakeholder representation in program development. In the case of North American mitigation strategies, this is reflected in the focus on national partnerships among key stakeholder groups. Finally, the normative undertaking of CTA, with its emphasis on symmetrical social learning, is evident in efforts aimed at promoting education and reflexive dialogue among a wide range of stakeholders through government-sponsored programs and industry forums.

While it is of course important to maintain a distinction between CTA as an approach to technology policy research and contemporary disaster mitigation strategies as a new policy approach to the management of critical infrastructure, the key point here is that their principles and aims display a high degree of resonance and similarity.

---

[3] Evident in this proposal is a systems-theoretic idea similar to that of homeostasis but more accurately based on a second-order cybernetic model that addresses the problem of evolving systems (Leydesdorff, 1996).

How to Analyze a Technology Project

While some discussion of CTA's three related undertakings—the analytic, the political and the normative—has been provided thus far, this book concentrates on the analytic domain both theoretically and empirically. While I will consider both the political and normative aspects of CTA throughout the following chapters, I will not deal with them in depth, as they are worthy of separate projects. In fact, results from this analytic study could contribute to a more extensive project dealing with political and normative aspects in the field of critical infrastructure management.

As I have already noted, a central concern in CTA is with 'path dependency' or 'lock-in,' which results when a long-term commitment of resources is made to a technology project. Path dependency represents a kind of momentum or soft determinism in technological systems that CTA practitioners seek to understand and influence at the early stage of design. Previous work done in the field of science and technology studies offers a socially informed theoretical account of technology dynamics that sheds light on the origins of the path dependency problem.

Historical case studies in the social construction of technology (SCOT) purport to offer a coherent perspective on technological innovation, drawing on constructivist principles that consider both social and technical forces in tandem, within a framework that treats both successful and unsuccessful projects as analytically symmetrical. In other words, the SCOT approach begins by assuming the relative validity or equal merit of all stakeholder perspectives at the earliest stages of a technology project (what I will describe later in this chapter as 'demand articulation' and 'problem formulation'). The SCOT approach is an effort to employ this fundamental, perhaps radical assumption to set the stage for studying the various social and technical forces that enter into play as a particular technology project is implemented, matures, disappears, or is otherwise abandoned (Bijker, 1995; Bijker and Law, 1994). From a methodological standpoint the SCOT approach also provides a set of guideposts and criteria for conducting symmetrical analysis of the social and technical forces at play—what Bijker has referred to in his work as 'sociological deconstruction.' This SCOT-based method is drawn from a synthesis of work undertaken in science and technology studies and it is helpful inasmuch as it is possible to apply it to perform empirical research not only for historical cases but also for contemporary technology projects, as I hope to demonstrate with the case of public safety telecommunications taken up in this book. Before advancing the SCOT-based method further, however, some additional conceptual work is necessary, particularly in the area of providing clarification on the process by which technology projects are transformed into working technical systems.

Basic Terms and Concepts

While SCOT-based studies typically involve technical artifacts, the basic approach is applicable to any effort that seeks to undertake symmetrical analysis of technological developments. Because of possible semantic confusion in referring to technical 'artifacts' when dealing with infrastructures that might be more properly considered large technical systems, I have adopted the umbrella term *technology projects* to refer to all open-ended, contingent, and indeterminate efforts to establish a working technical artifact or system.

The SCOT approach to studying technology projects is structured in two stages. In the primary stage, the researcher begins with a sociological deconstruction by identifying 'relevant social groups' and their relationship to a technology project. I will substitute relevant social groups with the term, *interested parties* because it corresponds more closely to the nomenclature of policy and regulation. In the second stage of this sociological deconstruction, the researcher then seeks to identify the multiple meanings that each interested party ascribes to the technology project under consideration. Within the SCOT literature, this is usually referred to as 'interpretative flexibility' (Bijker, 1995, p. 45, 73). In keeping with the idea that technology projects are open-ended and contingent undertakings, it seems to me that the term *problem formulations* might better characterize the multiple and indeterminate perspectives of the interested parties and more clearly reflect the uncertain domain of the quasi-subject in contrast to a view that regards design as an act of proposing solutions to pre-defined problems. At this very early stage of design it is important to recognize that it is the problem itself that often must be defined and that, in many instances, the interested parties may not be in agreement on that very fundamental question.

Problem formulations include a number of related elements of interest to the researcher. First, they embody a set of forces or actors, which may be termed 'human' or 'non-human' in character. Non-human actors can range from forces of nature through to legal and historical decisions that interested parties might deem to be relevant to the problem formulation. For example, in the case of a technology project involving wireless telecommunications infrastructures, some interested parties might place emphasis on spectrum allocation and legacy switching equipment as constraining factors in their problem formulation. While it is possible to argue that these elements are ultimately products of human action (as indeed they are) it is nevertheless important at this stage that researchers accept the interpretation as given by the parties themselves. As Latour has put it: 'How to frame a technological investigation? By sticking to the framework and the limits indicated by the interviewees themselves' (Latour, 1996, p.18). If interested parties give weight to non-human forces in their problem formulation, the researcher should treat this as a valid interpretation because it may indicate an important marker in an unfolding discourse among parties with different perspectives.

The second element of problem formulations is a set of unresolved issues that the interested parties have expressed as relevant to the technology project. This may be expressed as a form of *demand articulation* and describes the perceived need(s) for which a technology project might be undertaken (Van Den Ende *et al.*, 1998, p. 11). In terms of causal connections in the study of a technology project, it may the case that demand articulation precedes technical problem formulation, as in the call for a new technology, or it may be taken up in conjunction with problem formulation after a technology project is underway.

In the case of mature technology projects, sociological deconstruction is the method by which the researcher traces the developmental process from conception to subsequent closure or stabilization into what Bijker calls a 'socio-technical ensemble' (Bijker, 1995, p. 49, 84). In the case of immature or rapidly changing technology projects such ensembles may only exist as ideas put forth by interested parties as a result of demand articulation and problem formulation. Rather than as socio-technical ensembles, I refer to these configurations in their virtual form as *design propositions*. To put it more succinctly, design propositions emerge from problem formulations and represent attempts by interested parties to establish a specific socio-technical ensemble. It can thus be expected that if several interested parties have different problem formulations then a number of alternative design propositions will be put forward for consideration. These propositions may be complementary or competitive in one or more dimensions.

Stepping through the Analysis

In the original SCOT approach, the investigator begins with a finished product and works backward to provide a detailed account of the indeterminate technology project out of which it emerged. For such a constructivist-type analysis to be valid it is critical that the account of the technology project is symmetrical, meaning that it includes both successes and failures and does not adopt an *a priori* evaluation of the capacity of human and non-human actors to influence action. In its original formulation, however, the SCOT approach presents an obvious problem for Constructive Technology Assessment because it is based on a deconstruction that requires an existing artifact as basic source material. For example, Bijker's (1995) work examines artifacts such as the safety bicycle and fluorescent lighting, both of which achieved a significant degree of closure several decades previous to his analysis. How then does one undertake a study situated in the early design phase of such a technology project, as CTA would have us do? Under these circumstances the investigator likely has no stabilized artifact or system as a starting point for analysis. Is it then possible to use the SCOT approach if one *begins with unfinished technology projects*? The question again raises the problem of the Collingridge Dilemma: how do we anticipate the appropriate range of interested parties when a technology project is at its earliest phase of development? The question also raises a more fundamental concern for the investigator: how do we establish the very presence of a coherent technology project from what might be a widely scattered field of social groups articulating various demands and problem formulations?

The challenge presented by the Collingridge Dilemma is compounded if we accept the SCOT principle of symmetry and do not give priority to any prescribed set of interested parties at the outset of the investigation. Bijker emphasizes this point, claming '[it] is not a matter of postmodern relativism but of recognizing that there will always be other actors who contribute to the construction of society and technology, actors that cannot be controlled' (Bijker, 1995, p. 288). Because actors are hard to control, Bijker is pointing out that it is therefore impossible to anticipate which actors might become interested parties in a technology project. Given this seeming impossibility of placing boundaries around a technology project, how is the investigator to identify 'interested' parties and to establish which groups might represent the most important or interesting groups to study? Bijker himself suggests that the researcher ought to adopt the snowball method to identify relevant social groups. This approach may work to expand the potential range of interested parties through iteration but it has been criticized as 'inadequate for identifying unrecognized and missing participants, while its emphasis on groups overlooks social structures that might account for such absences' (Klein and Kleinman, 2002, p. 32). On this point, critics have noted that snowballing may induce bias into the investigation, either by leading researchers to follow a trajectory inadvertently determined by the early identification of interested parties, or moreover, by possibly overlooking systemic elements that may in fact be important forces of influence in shaping the views of those who are able to participate in a technology project.

In light of this critique, an investigator might then decide to turn to the social structures in which technology projects typically come to be articulated. Granted, this does not guarantee that all social groups will be represented but it does provide a starting point for identifying a core set of interested parties at the conceptual stage of development. Even without a stabilized socio-technical ensemble at hand, a technology project will serve as the basis for the SCOT approach, provided it is defined well enough to attract interested parties to a locus of interaction. Here the investigator can look to existing structural arrangements for such interaction and can establish a claim for the empirical existence of a technology project. Examples of such loci will include government-sponsored public forums, industry working groups, and regulatory hearings. Within these settings it may be possible for the investigator to identify those parties who have expressed interest or are likely to express interest in a technology project. Instances of this, such as calls for a regulatory hearing or a proposal for a standardization initiative, are usually instances of demand articulation and may therefore provide a source of evidence to establish the presence of a technology project.

Table 2.1 summarizes the modified SCOT approach, with demand articulation added in advance of sociological deconstruction (or what some investigators refer to as 'socio-technical mapping').

I have now presented the first two steps to a method suited to the analytic undertaking of Constructive Technology Assessment. The third step identifies various means by which interested parties come to be directly involved in a technology project. A theoretical account of the translation of problem formulations into design propositions serves as the basis for a 'layout of interventions' (Van Den Ende *et al.*, 1998, p. 9) that supports analysis of stakeholder participation and draws on complementary work done in Actor Network Theory.

Table 2.1 Steps in a constructivist analysis

Step	Objective
Demand articulation	Establish the existence of a technology project
Socio-technical mapping	Identify actors and issues using a thickly descriptive analysis of interactions
Layout of interventions	Evaluate the various strategies by which problem formulations are generated, translated into, and accepted (or rejected) as design propositions.

Translation and Alignment

Actor Network Theory (ANT) is another branch of study within the domain of science and technology studies that provides a perspective on the dynamics of innovation informed by the principle of symmetry.   Similar to the SCOT perspective, the ANT approach equates a successful technology project with a form of rhetorical achievement, in the sense that it involves the successful *alignment* of human and non-human actors to create a stabilized socio-technical system. Actor Network Theory, however, offers a more extensive theoretical perspective on the alignment process. Whereas SCOT only touches upon this process, ANT offers the concept of 'translation' to describe the process by which actors are drawn into a technology project.   The CTA literature sometimes refers to this process as 'modulation' while ANT writers have also described it as 'enrolment' or 'scripting' (Latour, 1995; Law, 1999). In view of recent work in design studies that has argued persuasively for a broad definition of rhetoric to include the design and development of technological artifacts, I would also suggest that the process could be considered a rhetorical process.[4]

In advance of translation, however, parties must acquire a *congruency* of interests, though not necessarily consensus on the meaning of a technology project (Leyten and Smits, 1996). Translation is a necessary prerequisite to any alignment of interests but it does guarantee it.   In order to move forward on a design proposition, interested parties must be bound by commitment with one another.

---

[4] Buchanan (1987, p. 26) has argued that design resides within the domain of invention and is thus properly associated with rhetoric as 'the study of how products come to be as vehicles of argument and persuasion about the desirable qualities of private and public life.' He presents a compelling case that claims 'the pattern of rhetoric in twentieth-century design builds on distinctions which were established early in the formation of rhetorical theory and developed to meet changing circumstances' (p. 44). He then suggests that the investigation of design in theory and practice centres around four themes closely related to the traditional divisions of rhetoric: invention and communication, judgment and construction, decision-making and strategic planning, evaluation and system integration (p. 45). Further study of Buchanan's work in this area could inform a more detailed study on how problem formulations are turned into effective design propositions.

Akrich describes this process as one of 'superposition' of representations in the design-development phase of a technology. A successful alignment is measured by the quality of relations among interested parties:

> As ... put forward in actor-network theory and shown for many cases, a new technological system will succeed only when it is able to attract a whole universe: a network of socio-technical relationships has to be put together, persuaded, and enlisted. In the final analysis, verifying the viability of the proposed combination of user representations entails determining if a system is able to relate harmoniously to appropriate networks, and ensuring that the various implications of the proposition are not conflicting and do not introduce intolerable stresses or constraints. (Akrich, 1995, p. 177)

In this formula 'viability' is equated with a successful alignment of interests and is measured by a high degree of commitment among interested parties when, as Akrich puts it, 'various implications of the proposition' are examined in light of anticipated consequences put forward by those involved in the assessment. Viability, in other words, is a relative measure closely linked to the composition of the group of stakeholders invited to participate in the design-development of a technology project. The inherent challenge to successful alignment in CTA is the expansion of conditions against which viability must be assessed, which means that the horizon of 'implications' must be expanded to include those views not previously addressed in earlier forms of technology assessment. As this horizon of implications is extended, previously overlooked stakeholders become involved and a concomitant shift in the criteria of viability takes place.

Successful alignment of interests is in some respects the quintessential moment of translation, when a set of diverse interests is 'purified' through a common framework of consensus (Latour, 1997). For alignment to occur, a congruent point of singularity that motivates and directs activity among the various interested parties must be created (Bijker, 1995, p. 276). Latour has termed this process *circulation through translation* (Latour, 1999), where interested parties come to recognize their self-interest in a technology project and orient their actions accordingly. In this formulation, the prime mover of any technology project may be compared to an autopoietic process that spawns second-order alignments among other actors and their associated networks.[5]

The perspective offered by Actor Network Theory is useful for the several discrete moments it suggests are involved in the movement from problem formulation to design proposition, and then to working system. First, the initial stage of translation suggests a requirement for both a space and a process of establishing legitimacy among interested parties. Second, congruency requires a means of establishing and maintaining a basic shared frame of meaning (not necessarily consensus) among interested parties with respect to a technology project. Third, successful alignment requires a vehicle (or what we might also call

---

[5] See, for instance, Luhmann (1995) for an autopoeitic theory of social systems.

a 'medium') of inscription—in other words, a means of firmly binding the actors together in commitment to a design proposition.

Requisite Conditions for Alignment

The constructivist approach typified by the triple alliance of CTA, SCOT and ANT helps to provide an account of the design nexus that supports an applied analytic framework for mitigation-oriented policy research. A very simple illustration of the relevance of this approach as a narrative framework for technology projects in action is found in the technical standardization process. The preliminary acceptance of the need for a standard is the principle step necessary to establish a fixed frame of reference for the circulation of interests among diverse actor networks. To achieve successful alignment around a specific standard, however, three key elements are necessary: a space or process of legitimacy for actors to come together; a means of establishing and maintaining congruency on the goal of the undertaking and the steps to be taken toward it; and a vehicle of inscription to lock parties into commitment with one another once some degree of consensus is achieved.

Similarly, conferring legitimacy on interested parties requires a space and/or process by which those interests can be expressed and pursued. Generally speaking, we can readily identify two kinds of spaces that provide for legitimization of interests: regulatory forums and voluntary forums. In terms of critical infrastructure, these forums can be further classified as either government- and industry-directed. Industry-directed forums tend to specialize in both constituency and focus, whereas government proceedings tend to provide an opportunity for a greater diversity of participants and focus on a wider range of general issues. Often there is no single forum or single combination of forums in which critical infrastructure concerns and interests can be exchanged among a broad constituency of suppliers and users of services. Technical groups tend to address important details at the engineering and physical network level, while users' groups are concerned with service concepts, rather than technical implementation.

Beyond the problem of establishing a space of legitimacy, the next concern is the formation of congruency around which to orient activities between interested parties. This is a particularly difficult problem insofar as CTA practitioners advocate an open and participatory process of symmetrical social learning. Practically speaking, this means bringing together engineers, service providers, equipment vendors, customers, government officials, and members of the general public to discuss ideas and concerns. The achievement of congruency is challenged by the disparities in expertise, motivation, and vision that reside within such a diverse group of participants. However, the basis for congruency in critical infrastructure may be formed around concerns such as national security/emergency preparedness, sustainable development, or regulatory parity across regions or between service sectors.

Finally, a vehicle of inscription is necessary to bind actors into commitment to each other. A number of methods are commonplace today such as legislation, sector specific regulations, technical standards, memoranda of understanding

(MOU), and various other forms of contracts. While each method differs in its means of compliance, all have in common the provision of an inscribed point of consensus that serves as the basis for diversified but coordinated action.

Intervention Strategies

Having identified the conditions necessary for involving actors in a technology project, it is equally important to consider how they might be combined with other policy instruments to influence the trajectory of a technology project. CTA practitioners have identified three 'generic' strategies commonly used by policymakers as a means to influence technological development in the early phases of a technology project: (1) technology forcing; (2) strategic niche management; and (3) loci for reflexivity (Schot, 1998; Schot and Rip, 1996). Taken together with the conditions for alignment, these strategies establish a layout of interventions suited to mitigation-oriented policy research.

Technology forcing is defined as a demand-side strategy where a social actor with some measure of authority, often a government or regulatory agency, stipulates desired impacts and directs technology development toward those ends through regulations or other incentives (Schot and Rip, 1996, p. 258). Technology forcing is a strategy that places emphasis on a vehicle of inscription for binding actors into alignments in order to achieve identified policy objectives. A common example of this strategy is the introduction of clean air legislation designed to influence technology development in the automobile industry (i.e., to force the development of lower emission technologies).

Strategic niche management is a supply-side strategy where industry stakeholders orchestrate product development 'through setting up a series of experimental settings (niches) in which actors learn about the design, user needs, cultural and political acceptance, and other aspects' (Schot and Rip, 1996, p. 261). Organizational learning is often emphasized in this strategy, with the objective of improved development processes and more precisely targeted technology products. This strategy places emphasis on generating congruency or shared frames of meaning among technology suppliers. An example of this strategy is the formation of a government-industry partnership to explore new market possibilities for an emerging technology project (e.g., alternative fuel systems).

The loci for reflexivity strategy is one that modulates interactions between supply and demand, attempting to 'create and exploit *loci*: actual spaces, and institutionalized linkages' between actor networks (Schot and Rip, 1996). Temporary loci, such as consensus conferences or special workshops, provide one opportunity to bring together supply-side and demand-side parties. Another approach is to create new loci or 'regular nexuses' or to modify the mandate of existing loci in order to ensure regular engagement between interested parties. This strategy for intervention places emphasis on a space of legitimacy as the means to achieve better actor network alignments. In other words, it is based on creating institutional forums that recognize a wide diversity of interested parties on both the supply and demand side, and could include third parties who might be concerned with certain externalities (e.g., pollution or other unwanted impacts) that might accompany the implementation of a technology project. Table 2.2 summarizes the basic intervention matrix and includes examples for each configuration.

Table 2.2 CTA intervention matrix

	Technology Forcing	Strategic Niche Management	Loci for reflexivity
Basis for legitimacy	Government legislation; policy; regulatory bodies.	Government research programs; private R&D investment and working groups.	Government or industry associations; public forums; congresses.
Congruency	Public welfare.	New markets.	Improved professional practice; knowledge sharing.
Vehicle of Inscription	Regulations; technical standards.	Partnerships and alliances; tax credits; direct government funding.	Memoranda of understanding; non-disclosure agreements.

Each strategy will exhibit different qualities related to the three conditions for stakeholder alignment. For instance, the basis for legitimacy of participation in each strategy may differ. In the case of the strategies of technology forcing and loci for reflexivity, the principle of democratic participation as embodied in government policy may provide the basis for wide public consultation in a technology project. This form of democratic participation is typical of regulatory hearings or consultations that deal with issues considered relevant to wider community interests and it may be a requirement prescribed by legislation or policy documents. In the case of strategic niche management, the basis for legitimacy might be more narrowly defined, as by membership within a specific industry or by demonstrating expertise or an investment commitment to a technology project. Under these conditions, the basis for legitimacy is not public participation but rather, that of experience or financial capability, which may be prescribed by regulations for government research funding, by exclusive membership in an association of interested parties, or by invitation to closed consultations.

The other two conditions for successful alignment, congruency and inscription, also differ across the three generic strategies for intervention. The principle of congruency—meaning the shared frame of reference among interested parties and again, not necessarily consensus—will likely vary according to the strategy. In the case of government-led technology forcing strategies, the principle of public welfare tends to provide a common frame of reference, whereas with strategic niche management, the common referent is central to the success of the strategy and is typically based on the creation and/or cultivation of new markets. A strategy based on loci for reflexivity may shift the principle of congruency to a common frame of reference involving improved business practices or knowledge sharing, such as one might find at professional conferences or academic congresses.

The vehicle of inscription will likewise differ according to the strategy being applied and the type of stakeholder alignment being sought. Technology forcing is a strategy that draws on an effective means of binding interested parties to conformity, possibly relying on coercive tactics such as regulations, reward structures based on targets, or relying on systemic measures by setting requirements for the implementation of specific technical standards. Strategic niche management may not require such formal measures but may seek to obtain alignment of parties through contractual arrangements in the form of alliances and partnerships. Government-based incentives such as tax credits or targeted funding programs may also provide a vehicle of inscription to encourage private initiatives in specific innovation activities. Loci for reflexivity are less reliant on developing or enforcing vehicles of inscription as the primary objective of the strategy but will nonetheless employ methods such as memoranda of understanding or non-disclosure agreements as a means of binding interested parties to mutual commitments.

While each intervention strategy is a unique combination of specific methods for initiating a technology project and for gaining alignment among interested parties, CTA practitioners have cautioned against the false assumption that each strategy is mutually exclusive. These are analytic distinctions that help policy researchers to more clearly understand the means by which intervention can take place in technology projects and they provide a basis by which such actions can be assessed for their effectiveness. The reality of any empirical technology project will likely be a complex combination of generic strategies that in itself is an important point of analysis as a kind of meta-strategy that can inform a multiple and layered approach to interventions involving a combination of forcing, niche management, and reflexivity. Throughout the remainder of the book I will refer to this intervention matrix as an analytical framework for assessing the role and feasibility of various strategies used by interested parties to influence growth and change in critical infrastructure.

Summary

Drawing out the importance of the design nexus as an essential site for the study of technology projects, this chapter introduced Constructive Technology Assessment as an approach suited to mitigation-oriented policy research. CTA is defined as a new design practice based on process-oriented principles and a constructivist theory of technology dynamics.

Using CTA as a general framework, I then presented a method for studying technology projects based on the 'sociological deconstruction' of SCOT. The framework was enriched further, through a theoretical account of the process by which interested parties come to be involved in a technology project. This account specifies three conditions necessary for the implementation of a technology project: (1) the need for an initial space/process of legitimacy among interested parties; (2) a principle of congruency that can form a common point of reference for participants; and (3) a vehicle of inscription to bind actors to commitment to one another.

I then combined these conditions with CTA's three generic intervention strategies of technology—forcing strategy; strategic niche management strategy; and loci for reflexivity strategy—to produce a basic intervention matrix that can be applied to study the process of growth and change in critical infrastructure. The next step toward this end is to provide an operational basis for the analytical framework by positioning the intervention matrix within the domain of networked infrastructure and large technical systems.

Chapter 3

Turning to the Empirical

The Dynamics of Large Technical Systems

Moving ahead with the intervention matrix presented in the previous chapter to the actual study of critical infrastructure requires a further step in conceptual development. This is because much of the constructivist literature described in the previous chapter and used for the development of the analytic framework was based on the study of discrete technical artifacts. For instance, Bijker's formative study of socio-technical change focused on three cases involving the development of the safety bicycle, bakelite (an early form of plastic), and fluorescent lighting (Bijker, 1995). Critical infrastructure, by contrast, involves what is more appropriately termed 'large technical systems' such as water supply, energy grids, integrated transportation networks, and telecommunications. These differ from discrete technical artifacts insofar as they are comprised of diffuse networks of interacting components and subsystems. Indeed, this is often what makes them 'critical' in the first place: the fact that they are so pervasive throughout modern societies means that everyday social and economic activities rely on them to a high degree. This important difference between artifacts and networks demands additional consideration and further conceptual development of our intervention matrix to account for the unique properties and difficulties associated with large technical systems.

The genesis of constructivist studies of large technical systems (LTS) is generally attributed to Hughes' seminal work on the development of electricity networks in Europe and America throughout the early part of the last century (Hughes, 1983, 1987; Mayntz and Hughes, 1988). While Hughes was interested in the historical process of development in large technical systems and provides a widely lauded contribution to the history of technology, his work represents much more insofar as he contributed the first systematic analysis of LTS development, introducing key concepts and a systems-based evolutionary model that has since spawned a distinct field of technology research. In one of the first major collections of work in the field of large technical systems, Hughes' influence on the field is characterized as that of introducing a socio-technical systems perspective, 'linking technical apparatus to the engineering systems, and in turn these to manifold organizational, economic and political actors and structures' (Joerges, 1988, p. 11). Given its emphasis on linking the technical with the social, Hughes' model is also typically associated with science and technology studies, including the Social Construction of Technology (SCOT) and Actor Network

Theory (ANT) approaches introduced in the previous chapter (Cutcliffe, 2000, p. 30).

Hughes' evolutionary model is based on observations drawn from historical research, indicating that 'large, modern technological systems seem to evolve in accordance with a loosely defined pattern consisting of several phases: invention; development; innovation; transfer; and growth, competition, and consolidation (Hughes, 1987, p. 56). Hughes points out that these are not strictly sequential phases but tend to overlap and backtrack; yet a pattern is discernible 'because of one or several of these activities predominating during the sequence of phases' (p. 57). The model is important on the one hand because it establishes the baseline for a wider field of related scholarship, yet on the other hand, it is controversial because it is a 'phase model' criticized for the pedigree of its historical approach to technology studies that may not be valid when it is applied to studies of contemporary technical systems. Expanding on this critique, one writer has presented three questions for consideration in assessing the suitability of Hughes' phase model for research projects that are concerned with more recent developments in large technical systems (Joerges, 1988, p. 15):

- Is the phase model compatible with evidence about the same systems when produced by disciplinary approaches other than history (e.g., political economy)?
- Can the model be generalized to the development of other types of LTS beyond the scope of Hughes' original study of electricity systems (e.g., transportation, communications)?
- Can the model be generalized to post-maturity stages in LTS development (e.g., expansion, upgrading, de-regulation)?

Joerges responds to his own questions, claiming that the phase model can be used to make sense of technological systems, first, by using historical evidence and second, by using data generated by other disciplinary approaches, provided the analysis remains focused on the link between technical and social factors. On the second question, however, Joerges' view is that empirical case evidence suggests that the Hughes model may not be applicable to other types of large technical systems, especially 'in the case of implantation of new subsystems in old, "mature" LTS' (1988, p. 15). One example of such a case can be found in the introduction of a communications control system within a large transportation system—a development particularly relevant to the interdependency dilemma. The third question remains unanswered in Joerges' paper and as I interpret it, seems to query the comprehensiveness and capacity of Hughes' model to account for the latter phases of growth and change once a large technical system continues to evolve beyond what might have been initially considered its mature stage of development. This question might also be restated in the obverse, to ask what kinds of events in the 'late maturity' phase of development might disrupt previously established stability and momentum, leading to renewed efforts at system building? And how might this subsequent period of system building correspond to Hughes' phase model? These questions are especially relevant in light of the advent of the regulatory reforms and the introduction of new digital technologies that are continuing to transform telecommunications and other forms of critical infrastructure. This unanswered question raises the matter of the validity of Hughes' phase model for understanding contemporary developments in large technical systems. Further, the matter begs the question of whether or not such developments reflect a changing set of socio-technical relations.

Difficulties in Applying the Phase Model

Despite the potential problem of applying it to contemporary developments, Hughes' phase model does offer an important contribution to the study of critical infrastructure: namely the proposition that large technical systems exhibit various phases of development and that each phase tends to exhibit significant differences in stakeholder involvement. This observation suggests that intervention strategies appropriate in one instance may not necessarily be effective at different stages of development either within or across the domain of different types of infrastructure. For the purposes herein, while such a consideration is important, it is less immediately relevant than the methodological problems it implies. In attempting to position a large technical system within a historical phase of development, the researcher is confronted with at least three obstacles. First, any attempt to locate a large technical system at a specific phase of development will need to account for cases of uneven development across that system. This leads to a second problem of establishing boundaries around what may be highly interdependent and geographically dispersed systems. The researcher will also need to account for the effects of technological innovations and other disruptive forces on a large technical system, as they may surface from beyond the apparent boundaries and control of the dominant institutional actors.

With respect to the first problem, is it realistic to claim that a large technical system such as the public information infrastructure is evenly developed across most countries? With current growth and change, are we witnessing a single, unified pattern of development or something more akin to a set of related but uneven developments across regionally related subsystems? Hughes' concept of the 'reverse salient' may be applicable here, where the problem of uneven development is taken into account as 'technical or organizational anomalies resulting from uneven elaboration or evolution of a system' (Joerges, 1988, p. 13). By adopting this approach, a large technical system can be analyzed as a single unit of analysis with an advancing front but with less developed components operating within the system. While the notion of the reverse salient may address the problem of uneven development, the problem of establishing boundaries for the analysis remains unresolved. For instance, what constitutes the edges of a networked infrastructure such as that of a telecommunications system? Should it be conceived of as a single system, as a composite of several national or regionally competing service providers, or as a set of functionally distinct but physically connected (e.g., voice versus data services) networks? Or for that matter, is it more accurately defined as a subsystem within a continental or global-wide infrastructure system? In order to undertake a study of growth and change in such a system, the researcher is obliged to specify the nature and extent of the system under study, accepting that such specification will determine in part the technical and institutional characteristics that might be attributed to that system.

Some have criticized Hughes' phase model for under-emphasizing the importance of external factors on system development, raising the question of how we might conceptualize disruptive forces and other unforeseen influences. Werle (1998), for example, argues that Hughes' model confronts recent social and technical developments, which appear to indicate a qualitative re-configuration of these systems, driven by influences from beyond their boundaries. Characteristic changes that Werle has observed include a shift from monopoly to competitive structures; a movement from centralized to de-centralized control of system elements; and a shift from tight coupling of system elements to looser federations of organizations and elements. In effect, he argues that it may be difficult for Hughes' model to account for the complexity of influences on contemporary large technical systems.

To illustrate his point, Werle characterizes Hughes' model as one that assumes a more or less hierarchical view of development based on the governing power of systems-builders. This view, he argues, blinds analysis to a richer understanding of the multifaceted influences that shape growth and change in contemporary systems. He claims that 'looking at the development of technical networks from the angle of governance forms suggests *a one-way relation* between [system-builders] and technology' [emphasis added]. 'Co-evolution,' he claims, is a more accurate conceptualization in that it takes into account the reciprocal influence(s) of path dependency and other variables in shaping the development of a large technical system. Werle gives evidence of such co-evolutionary dynamics in his account of the growth of the Internet, where he seeks to demonstrate how the hierarchical governance structure of the monopoly-era regime drove the growth of proprietary data networks in the 1970s and 1980s. The subsequent fragmentation of data networks in turn drove the development of TCP/IP and other 'bearer services' toward integration. Today, the success of TCP/IP has resulted in the explosive growth of the Internet and other data services, setting off new debates about appropriate governance structures and provoking significant changes in telecommunications policy and regulation, such as that evident in the case of Voice over IP services (e.g., Mindel and Sirbu, 2000).

In many respects the monopoly-era regime seems to fit well with Hughes' model, when state-controlled operators and private monopolies served as primary systems-builders with vertical control over equipment and services, thereby acting as key influences on the development of these large technical systems. Yet, as Werle notes, this form of governance is partly responsible for creating a situation in which independent developers of data networks were excluded from possible interconnection with the public switched telephone network, thereby making 'it unlikely that the local, regional and corporate networks for voice and data communications would be integrated into one encompassing network' (Werle, 1998). As a result, these competing independent firms found themselves in a world of increasingly proprietary and fragmented data networks. Over time this forced the development of new technologies and services that provided a soft integration of proprietary networks through protocols like TCP/IP, which began to create an increased demand for new data services. Soon the incumbent monopoly carriers found themselves confronted by the very forces that had been initially excluded from entering into competition with them. Werle's observations offer an insightful example of what might be termed a *forward salient*, where an existing regime of control provokes outside developments that later return to out-flank that same governance structure. His point is that contemporary developments in large technical systems, and particularly those involving information infrastructures, must be appreciated as a product of co-evolving forces and not as a product of the sheer willpower of designated systems-builders. With this we are again confronted with the problem of establishing boundaries for the analysis of large technical systems.

Defining a Large Technical System

Given the challenges I have just outlined, how is it possible to delimit and effectively analyze a rapidly changing technical system such as the public information infrastructure? What in fact *is* a large technical system? In his attempt to establish a definition, Joerges (1988, p. 21) offers three approaches, each of which I contend is rather problematic when applied to networked communications systems. First he says we might look at the size of the organizations that control the system. A *large* technical system is one that *large* corporations control. On these terms, a telecommunications network might actually be considered several interconnected systems operated by a number of large competing carriers. A second approach, according to Joerges, might focus on the externalities produced by or under the influence of a technical system. Here the emphasis is on unintended social or ecological effects of the system, where *large* is often equated with *high risk,* as in the case of a large technical system that produces and distributes nuclear energy. This approach suggests extending or perhaps erasing the boundaries of the technical system itself and integrating it into a wider social and ecological web of effects and interdependencies.

The problem with both approaches is the confusion or blurring of systems with their operating environments (Hughes, 1987, p. 53). Analytically, these two approaches lose sight of those 'features of large-scale technical systems which should be kept separate [from their environment] in order to explain their [unique] dynamics' (Joerges, 1988, p. 22). This is an important consideration when policy research is specifically interested in the process of infrastructure development. In response, Joerges advises that researchers maintain 'close(r) reference to the scale of the technical core—both materially and otherwise—of LTS [in order to] give room to inquiries about the conditions and consequences of LTS' momentum' (Joerges, 1988, p. 22). In other words, if we fail to distinguish between a large technical system (LTS) and those forces that support it or are impacted by it, then we may inadvertently reduce our ability to analytically distinguish those attributes explainable by the significant properties of the system itself.

Joerges (1988, p. 23) therefore proposes a third approach that combines elements of the first two and provides for the 'preliminary delineation' of a large technical system in view of the fact that he claims that a more systematic, empirically-grounded definition is still wanting. His definition begins from an engineering standpoint, identifying technical systems with '...systems of machineries and freestanding structures performing, more or less reliably and predictably, complex standardized operations by virtue of being integrated with other social processes, governed and legitimated by formal, knowledge-intensive, impersonal rationalities' (Joerges, 1988, p. 24). He refers to this as a 'formal rationality' (p. 19) that captures the essence of a technical system and he advises that we might then draw the measure of scale (large or small) based on the quantity of activities 'materialized in such systems' as well as drawing the scale of other social processes (e.g., financing, regulation, etc.) necessitated by these activities in order for the system to function. In order words, he says we must first have a rationally designed technical system and then we are to define it as 'large' based on quantitative terms. Of course this formulation is equally unsatisfactory because it leaves us with the problem of making arbitrary distinctions of size. It also leaves the boundary problem unresolved. Joerges does acknowledge these difficulties, noting that such an approach ultimately requires an empirical knowledge of engineering and admittedly, 'much preliminary and arbitrary classification of qualitatively different technical systems' (Joerges, 1988, p. 24). Yet, he claims that this approach nonetheless provides some guidelines to distinguish certain types of technical systems that 'can be singled out as indisputably large,' such as:

- Those that are materially integrated or 'coupled' over large spans of space and time
- Those that provide support for the functioning of very large numbers of other technical systems, 'whose organizations they thereby link.'

Based on these two characteristics, typical large technical systems according to Joerges (p. 24-25) will include most critical infrastructure: integrated transport systems; telecom systems; water supply systems; and some energy systems. While some appliances may be connected to one or more supply systems (e.g., a dishwasher connects with the hydro grid and municipal water systems), these elements appear to reside at the outer edge of what one might consider to be the core of the large technical system proper. It seems that there may reside here a potential boundary condition drawn on a distinction between a core network and the terminal equipment that is attached to it. Yet, as Joerges himself admits, this kind of boundary setting becomes especially fuzzy when we think about connecting devices to a communications network, such as a personal computer, a private branch exchange, or a wireless local loop. Does a peripheral device or sub-network, become an integral component of the larger system when it is connected to other devices by means of a communications link? Is a communications link qualitatively different from other material couplings such as that of water or electricity? According to Joerges (1988, p. 25) a communications link *is different* from other types of connections, which means that the sort of 'material coupling' used to define a large technical system is also a significant consideration when setting boundaries; however, this important matter is not elaborated upon in the literature. In consequence, and en-route to solving this problem of setting boundary conditions, one must turn to the field of network economics to examine the notion of 'network externalities' for its relevance to material coupling to consider how it might better account for differences between communications networks and other types of large technical systems.

Network Externalities and Complementarity

The field of network economics provides a useful distinction between standalone artifacts and networks to further extend the notion of material coupling as it applies to Joerges' definition of a large technical system. Shy (2001), for instance, has observed that while many consumer goods are available as standalone items, certain kinds of goods must be consumed in conjunction with other products or services. A loaf of bread, for example, can be consumed by itself without an active interface to a network of components (that said, a wide range of support systems is required to get bread to your dinner table). A computer or CD player, however, requires a body of software to provide it with value in the marketplace. Similarly, a telephone or modem requires an active voice or data network to provide it with value. In fact, the more potential connections available over the network, the more value that is likely to accrue to the service. While the *cost* of some services such as electricity or water supply may drop in proportion to the number of users of the system, the functional value of the service to each customer does not necessarily improve and may in fact decrease with loading. With communications networks, however, larger numbers of connections means an increase in the functional value of the network because more potential points of contact are made available to all customers simultaneously. This is the notion of *network externalities* and marks a qualitative difference between the large technical systems that support dishwashers and those that support networked computers or telephones.

Network externalities are achieved through *complementarity*, which is a concept also found in the network economics literature (Shy, 2001, p. 2) and further helps to illuminate Joerges' assertion that large technical systems are 'materially integrated ... over large spans of space and time.' In order to produce network externalities, each of the nodes and links in a technical system must operate together to provide efficient connections between the various end-points. Complementarity describes this arrangement and is taken to be a function of two related operations—one technical, another social. The first of these operations is *compatibility*, which refers to coordination of design specifications based on mutually accepted standards. Compatibility, or 'interoperability' as it is sometimes called, is the technical dimension of material coupling needed to achieve network externalities. The second operation is *interconnection*, which refers to the regulatory, business, or other institutional arrangements needed to achieve a flow of services across a network. Interconnection requires compatibility as a prerequisite and provides the social dimension of complementarity needed to achieve network externalities.

According to one of the leading scholars in the field, the 'crucial relationship' in network economics is the compatibility of components, which tends to be grounded in technical standards as a key vehicle of inscription:

> ... for many complex products, actual complementarity can be achieved only through the adherence to specific technical compatibility standards. Thus, many providers of network or vertically-related goods have the option of making their products partially or fully incompatible with components produced by other firms. This can be done through the creation of proprietary designs or the outright exclusion or refusal to interconnect with some firms. (Economides, 1996, p. 4)

In conjunction with the fundamental importance of compatibility at the technical level, the ongoing process of regulatory reform around the world has also provoked an intense interest in the socio-economic dimension of network interconnection. In the field of telecommunications, this interest has inspired a number of studies that examine compatibility as a strategic socio-economic matter in the design of networks. Among these, several approach compatibility from within the field of large technical systems (see articles in Mayntz and Hughes, 1988; Summerton, 1994b). Other studies have explicitly identified design as a political economic force behind interconnection arrangements (Mansell, 1999; Mansell and Silverstone, 1996). Evident across these various studies is a similar concern among contemporary telecom regulators and scholars in the field of network economics:

> In a network where complementarity as well as substitute links are owned by different firms, the questions of interconnection, compatibility, interoperability, and coordination of quality of services become of paramount importance. (Economides, 1996, p. 6)

While network economists use the term complementarity, those writing about telecom reform tend to use the term 'interconnection' to embody the same idea, including both technical and social considerations. Interconnection is regarded by telecom analysts as the fundamental enabler of the future 'network of networks' (Noam, 2001) or as the 'cornerstone of competition' as Melody has termed it:

> Interconnection is fundamentally important because the telecom system must function as a single system. Users desire end-to-end services within an apparently 'seamless' communication network. They want connectedness and connectability. They do not usually care who own what facilities in the overall system, or how the communication links are established. As traditional telecom networks have grown from national to global dimensions and have been expanded to include competitive suppliers and new services, interconnection has become the key to defining the limits of telecom service networks and the structure of competition that can prevail in supplying them. (Melody, 1997, p. 53)

Interconnection has thus served historically as a regulatory instrument directly linked to telecom policy objectives, as both Melody and others have argued. Noam (Noam, 2001, p. 257), for instance, writes of three phases of interconnection policy that have guided the history of telecom development for much of the past century. Interconnection policy has undergone significant changes as political and economic interests shifted from pro-incumbent to pro-competition and then to market control strategies. The first phase was dominant in the previous era of regulated or state-owned monopoly telecommunications, where interconnection to the network infrastructure was highly restricted and open to only a few privileged operators and suppliers. In the second phase, interconnection policy was aimed at prying open these networks to permit competitive entrants, while in the third and current phase it has focused on actively promoting competition by facilitating innovation and entry through favourable interconnection arrangements.

Existing International telecom reform initiatives indicate that modern telecommunications systems may have indeed entered a new phase characterized by considerable emphasis on network complementarity. Recalling Hughes' phase model, it may be reasonable to argue that many large technical systems achieved an initial 'mature' phase in their development under monopoly regimes. General consensus in the literature on telecom reform, however, signals a renewed effort at system building with the advent of digital technologies and a concomitant push on several fronts from business users, computer industry, and regional economic development organizations alike as part of a co-evolving impetus to liberalize entry and to promote innovation and greater efficiency in telecommunications systems around the world (Davies, 1994; Melody, 1999; Winseck, 1998). Together, these forces have encouraged policymakers to shift away from direct intervention and toward a greater emphasis on what could be termed 'complementarity oversight' in the regulation of telecommunications and other types of infrastructure.

With this understanding of complementarity as a key feature of large communications systems, it is possible to work toward a resolution to the problem of setting boundary conditions. If communications networks are a unique form of large technical systems because of their peculiar network externalities, then the functional boundary of a telecom system is linked to complementarity, in both its technical and social dimensions. Where complementarity ends so does the network. A logical boundary for any telecommunications system is therefore the extent to which such complementarity can be influenced and enforced by a recognized authority. In effect, this implies two levels of analysis for consideration. The first is the extent to which such an authority can exercise influence over technical compatibility. The second is the extent to which such an authority can exercise influence over socio-economic matters of interconnection. Of note here is that the boundary of a national or regional telecommunications network is therefore not necessarily at the physical edges of the system but rather within the span of control over complementarity that enables a large technical system to generate network externalities. The boundary conditions of such a system are therefore largely political boundaries stemming from jurisdictional influence on the design of the infrastructure.

Network externalities are a unique feature of telecommunications systems, producing value for users of these systems by providing symmetrical access to other users of the system. Complementarity or interconnection is necessary to achieve network externalities. But to what extent does interconnection influence the design of the public information infrastructure and how might it be used as a regulatory instrument to influence the development of large technical systems more generally?

Arnbak's Functional Systems Model

The emphasis on the importance of interconnection as a regulatory instrument is addressed in the telecom policy literature by Arnbak (1997) who introduces a 'functional systems model' to differentiate between various forms of interconnection arrangements. It is based loosely on the Open System Interconnection (OSI) reference model, which was originally created to provide a standardized basis on which to design layered communications systems. Arnbak's model consists of four layers arranged bottom to top, from predominantly physical systems based on hardware elements to more logical systems based on software applications. Variations on this basic model have appeared in recent years but most bear the same key feature of using a series of functional layers to describe modern communications systems (Frieden, 2002; Sicker, 2002; Werbach, 2000). The primary layer is that of physical transport, which includes the provision of transmission capacity and basic physical interfaces. The second layer consists of network services, which include the provision of routing and gateway services. At the third layer lie value-added services, those applications and software platforms that provide support for the delivery of information content to a client. Finally, at the fourth layer reside information services where content is created, stored, and supplied to the client (Arnbak, 1997, p. 76). Table 3.1 summarizes Arnbak's Functional Systems Model.

Table 3.1 Arnbak's Functional Systems Model

Layer	Function
4	Information Services
3	Value-added Services
2	Network Services
1	Physical Transport

For a relatively simple example of Arnbak's functional systems model in action, we might consider the delivery of a weather bulletin to a mobile phone through a short message service (SMS).[1] A combination of wireline and wireless infrastructure (layer one) must enable end-to-end connectivity between a content provider and a handheld wireless device. Network services (layer two) then enable the correct routing (and perhaps billing) for the data from the content provider's network, perhaps through a series of gateways, on to an intermediary public network and eventually to the appropriate wireless service provider and the correct cell-site location (i.e., the 'local loop') for radio transmission to the wireless terminal that has requested the information.

In this scenario, a variety of suppliers may be required to support this simple service. At layer one, a physical transport operator is required to provide end-to-end connectivity. In some cases where large distances or organizational boundaries are crossed, several layer one operators may be involved in the physical delivery of the data through routers and other software elements (layer two). The wireless carrier, or perhaps a third-party service provider, must operate an SMS portal (layer three) that enables access to weather bulletins for its customers by creating a special user profile account. Finally, a content provider (e.g., a government meteorological agency) must supply weather data either in a raw or customized form (layer four). All layers must be interconnected in order to establish a working telecommunications service.

Interconnection Space

Policy research under these conditions can be extremely complicated, as numerous stakeholders may be involved in one or more layers of the functional model. Moreover, some of these functions may be regulated while others remain unregulated. In the case of the wireless weather bulletin, physical transport (layer one) includes the regulated domain of radio licensing, but content provision (layer four) may operate in an unregulated domain. Recognizing this complex environment, Arnbak uses the functional systems model to introduce the related concept of *interconnection space* as a means of modelling the multidimensionality of modern telecommunications networks:

> The policy issues facing a regulator are to decide to what extent the costs and benefits of a particular path through the interconnection space can be discovered—and allocated—by a free market, or if regulations are required to enforce desirable interconnection paths. (Arnbak, 1997, p. 79)

This notion of interconnection space provides a third axis and a further operational dimension to the intervention matrix developed in the previous chapter.. Figure 3.1 depicts the intervention matrix with the third dimension of interconnection space added to it.

---

[1] For an example of a similar type of service see the Cellular Emergency Alert Systems Association (CEASA) website at http://www.ceasa-int.org/

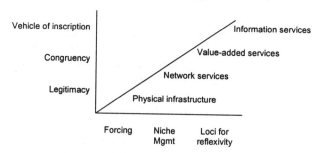

Figure 3.1 CTA intervention matrix with interconnection space

Designing Networks: Unbundling, Inwards and Upwards

The next step in adapting Arnbak's interconnection space to a study of critical infrastructure is to examine it in light of contemporary trends and related policy issues in order to identify areas of possible interest. To begin, interconnection within each layer can be of two essential types: those at the end of a network or those at a point between networks (Melody, 1997, p. 58-60). In other words, it is possible to distinguish between interconnection at the periphery or at the core of a network infrastructure. Furthermore, in the recent history of telecom reform, competition first emerged at the periphery of the network with the deregulation of customer premises equipment and has gradually moved toward the core elements, with liberalization of transmission and switching markets. This means that interconnection issues have tended to parallel this inward migration, yet for much of the recent history, the predominant issues have been confined to the lower layers of interconnection space, involving primarily physical transport and network services.

Melody has predicted that 'the more significant interconnection issues for the longer term relate to VAS [value-added services], resale and network management' because opportunities in these markets 'are essential to unleashing the new services … that will provide a foundation for the development of [future] information societies' (1997, p. 62). In effect, this suggests that while regulatory reform has increasingly resulted in deeper unbundling of network elements as it moved from the periphery to the network core, strategic business activities in the sector are now migrating toward upper layer activities of interconnection space in the form of value-added services (layer three) and information content (layer four). In many cases, these upper layers remain largely unregulated and therefore present a unique opportunity for innovation, assuming that potential developers and providers of such value-added services are granted interconnection to the public information infrastructure on fair terms and conditions.

Modular Networks?

The implications of such network unbundling may have a significant impact on future interconnection policy and could therefore come to play a significant role in the management of critical infrastructure. Noam, for instance, writes of three phases of telecom reform that involve a gradual evolution of interconnection design regimes toward what he terms the 'modularity paradigm.' In the so-called pro-incumbent regime of monopoly control, Noam notes that networks were tightly integrated in vertical structures under limited control. Since the advent of telecom reform and market-oriented interconnection policies, regulators have increasingly permitted horizontal linkages across all layers of interconnection space, which has resulted in the unbundling of network and service elements to support competitive access requirements. As this process continues to evolve, its logical trajectory of development is toward what Noam has termed a 'modularity' regime where 'service providers connect several modules together, or replace some network modules and interact with others' (Noam, 2001, p. 249). This modularity model posits a trend that is significant for the management of critical infrastructure: namely, the emergence of third party value-added service organizations or 'system integrators' that seek and combine physical and logical network elements to create customized service packages for clients.

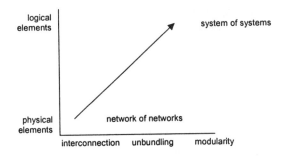

Figure 3.2 Noam's portrayal of network evolution

The modularity scenario is predicated on two developments that Noam (2001, p. 251) believes characterize the long-range impact of regulatory reform on communications networks. The first of these is a shift away from physical or hardware-defined network elements to more logical or software-defined network elements. The second shift reflects the gradual movement away from the simple interconnection of vertically integrated networks toward customized services based on a modularity paradigm consisting of many competitive network elements extended across all layers of interconnection space. Third parties then pick and choose from these elements to create functional service modules tailored to specific client needs. Figure 3.2 depicts an interpretation of Noam's portrayal of infrastructure evolution.

An implication of this trend similar to one presented by Melody is the suggestion that innovation activities will migrate to higher layers of interconnection and toward value-added services in the form of software-defined network elements. Noam claims that this trend will lead traditional carriers into 'uncharted waters' as other forms of expertise—such as those in the data networking and content provider fields—become increasingly prevalent (Noam, 2001, p. 252). On the surface, Noam's modularity model appears to present an opportunity for extensive market competition across the spectrum of interconnection space, and in a world of perfect competition, all layers and elements would be unbundled and open for new entrants to innovate and compete with one another. In its basic form, however, the model overlooks the fact that certain network elements—especially those with strategic value to incumbent and large operators—will likely require special treatment to ensure that public policy objectives can be accommodated.

To accompanying these observations about future developments in the public information infrastructure, Melody (1997, p. 59-60) also introduces four 'substantive issues' that are likely to be of continued relevance to interconnection policy. The first is the technical matter of standardization and interoperability that he claims is relatively easy to solve, although this may remain a critical point for intervention in network design, as will be discussed in the subsequent chapter. Second is the issue of defining terms and conditions for services that interconnection will support. Third is the competitive issue of access to markets and services, which includes establishing points of interconnection, levels and grades of service, and prices. Finally, the fourth substantive issue is that of regulation and the incorporation of broad social and economic policy into interconnection negotiations. Traditional examples here include universal service provisions, information society objectives, privacy and security.

Interconnection provides an operational basis for studying growth and change in the public information infrastructure. Arnbak's functional systems model and the related idea of interconnection space provide a means of specifying four layers or functional dimensions of interconnection. This establishes a third dimension to the intervention matrix, giving it greater precision. Noam's modularity model signals the growing importance of the upper layers of interconnection space, while Melody provides a number of issues that permeate interconnection and related design discussions.

Toward an Empirical Program of Research

I have argued thus far that a program of mitigation-oriented policy research should have two objectives: first, to understand the process of growth and change in critical infrastructure; and second, to assess the various means by which to intervene in that process in order to reconcile wider public interest with a program of long-term risk reduction, in accordance with the growing emphasis on national disaster mitigation strategies in various countries and regions around the world.

Toward this end, in the previous chapter I presented an analytic framework that provides a foundation for studying developments within critical infrastructures and moreover, that identifies a set of common intervention strategies capable of influencing the design of large technical systems. As illustrated in the previous chapter, the design nexus is a critical area of focus for policy research because it is the point where human intentions encounter institutional decision-making processes and where problem formulations are translated into design propositions for technology projects. The intervention matrix is offered as a basic intellectual tool to help researchers understand *in a more systematic way* the function and influence of the design nexus on community development—in this case, the development of the 'virtual' community of interested parties relevant to the management of critical infrastructure.

The intervention matrix is not enough, however. If it is to be an effective research tool it must be capable of supporting empirical assessments in specific case studies. Such assessments are necessary too, as a primary task for policy research in this field today is to better understand how growth and change happen in the complex and rapidly changing domain of critical infrastructure. With ongoing market liberalization and regulatory reform, the continuous appearance of new technologies on the horizon, and the context of important political (the threat of terrorism) and ecological (climate change) adjustments, it has become apparent that policymakers face an awesome challenge in their efforts to develop a coherent, integrated understanding of how complicated technical systems are evolving. The public policy research agenda should therefore begin with a coordinated approach to the study of specific instances of growth and change by employing a relatively standardized method that can lead, over time, to more generalized conclusions about how critical infrastructures evolve in a post-monopoly, liberalized, and innovation-driven world. By combining conclusions from studies done within and across various forms of critical infrastructure, our chances of designing and implementing effective mitigation-oriented public policy interventions will increase dramatically.

A discussion on the nature of large technical systems—or more specifically, of interconnection as a key point of analysis in the study of communications system— was introduced to advance toward the conduct of empirical research in this area. The next section in this book continues in this direction with three chapters reporting on a case study into what I have termed 'public safety telecommunications.' Stated in another way, the case study was an empirical assessment of the management of technological change within the public information infrastructure: namely the introduction of wireless enhanced 9-1-1 ('Wireless E9-1-1') service into the Canadian public switched telephone network. Now being introduced throughout North America, Wireless E9-1-1 is a public safety service that provides a location-enhancement function for mobile phone subscribers when they use their phone to call for emergency assistance.[2] Similar versions of the service are being considered for deployment in Europe and

---

[2] Similar proceedings are now underway to discuss the need for 'E9-1-1' for Voice over IP services that are beginning to gain a foothold in the consumer and business telecom sectors.

Australia. The Wireless E9-1-1 service is suited to mitigation-oriented policy research in the management of critical infrastructure in part because it represents a new service concept being introduced into the public information infrastructure, with significant consequences for the design of that infrastructure.

Wireless E9-1-1 is a large-scale technology project involving numerous interested parties—including incumbent telephone companies, wireless service providers, public safety agencies, equipment manufacturers, technical standards bodies, consumer groups, and national regulatory agencies—and offering multiple problem formulations and competing design propositions. Furthermore, the process of designing and deploying Wireless E9-1-1 is a contemporary development, placing in high relief many of the current dynamic pressures affecting the public information infrastructure, including the growing importance of the wireless sector, the convergence of digital systems and the arrival of location-based services, as well as issues of regulatory reform and competition policy. During my period of investigation between 2001 and 2003, Wireless E9-1-1 deployment had only recently commenced in North America, which also meant that, in theory at least, it had yet to achieve a stable state of socio-technical closure (to use Bijker's term) and was therefore amenable to symmetrical analysis.

Perhaps more to the point, however, the issue of public safety telecommunications is very much in the public interest, which means that Wireless E9-1-1 represented an opportunity to examine how public policy might influence growth and change in the public information infrastructure in accordance with principles set out in a mitigation-oriented policy framework such as Canada's National Disaster Mitigation Strategy.

Telecommunications: Intervention in the Public Interest

Telecommunications falls under federal jurisdiction in Canada and is therefore within the scope of the National Disaster Mitigation Strategy. The Ministry of Industry ('Industry Canada') has the primary role of presiding over telecommunications policy, including spectrum management and the Canadian Radio-television and Telecommunications Commission (CRTC), the arms-length regulator for the sector. During the period of investigation between 2001 and 2003, Canada's telecommunications infrastructure consisted of a small number of regional incumbent wireline telephone companies, a handful of independent telephone companies (most in the provinces of Ontario and Quebec), four national wireless service providers, and number of competitive long-distance and local exchange carriers. The introduction of the 1993 *Telecommunications Act* has made reliance on market forces a principal policy objective; however, the regulatory environment is something of a patchwork as the CRTC continues to regulate facilities-based carriers (those owning or operating transmission facilities), while non-dominant carriers are largely exempt from regulation (e.g., wireless service providers), as are re-sellers and other competitive service providers (see Grieve, 2000).

Section seven of the *Telecommunications Act* (Canada Department of Justice, 2001) establishes the overall policy objectives for Canada's telecommunications infrastructure and provides a helpful focal point for mitigation-oriented policy research concerning the public information infrastructure:

> It is hereby affirmed that telecommunications performs an essential role in the maintenance of Canada's identity and sovereignty and that the Canadian telecommunications policy has as its objectives
>
> (a) to facilitate the orderly development through Canada of a telecommunications system that serves to safeguard, enrich and strengthen the social and economic fabric of Canada and its regions; ...
>
> (f) to foster increased reliance on market forces for the provision of telecommunications services and to ensure that regulation, where required, is efficient and effective; ...
>
> (i) to respond to the economic and social requirements of users of telecommunications services;

The wording of the legislation suggests a set of three observations relevant to Canada's National Disaster Mitigation Strategy and to mitigation-oriented policy frameworks more generally. These observations also introduce a set of corresponding questions that were formative for the case study on public safety telecommunications, though they could be applied with equal relevance to any type of critical infrastructure.

First, section 7(a) implies the need to ensure the 'orderly development' of the telecommunications infrastructure to the extent that it is capable of supporting the broader public policy objectives related to a mitigation strategy (i.e., 'safeguarding' the public). In the face of rapid technological change and market liberalization, this suggests that some kind of regulatory supervision or *technology forcing* may be required to ensure that the telecommunications infrastructure will evolve in a coordinated way and according to established best practices that seek to reduce social risk and vulnerability. Is such supervision necessary? Is this supervision currently in place? If so, can it be improved? If not, what are the best means to implement it, given current policy and regulatory considerations?

Second, section 7(f) calls for an increased reliance on market forces for service provision, implying that initiatives taken up within a mitigation strategy should be fostered by competition wherever feasible, suggesting a role for *strategic niche management*. In terms of infrastructure development, this means a search for policy instruments to promote competitive research and development and to encourage the timely deployment of new value-added services aimed at reducing social risk and vulnerability. It might also mean regulatory intervention to ensure access to certain strategic network elements and services that will enable innovative services in support of mitigation-related activities. What policy instruments are available to support research, development, and technology transfer for the management of critical infrastructure? What are the key bottlenecks and strategic network elements that might influence and/or inhibit the deployment of new services?

Third, section 7(i) requires that the social and economic requirements of user communities be considered in the development of the telecommunications system. In terms of a mitigation strategy, this suggests a need to ensure stakeholder consultation is undertaken at critical stages of infrastructure development. Current stakeholder development may need to be expanded, or new processes may need to be introduced in order to better facilitate participation in critical decisions. This suggests that existing and perhaps newly formed *loci for reflexivity* may play an important role in a mitigation strategy. What then, are the important issues, challenges, and opportunities for expanding stakeholder participation in telecommunications infrastructure planning and development?

Telecommunications legislation in other countries is admittedly different in much of its wording and intent but in many cases the general policy objectives are similar, especially among Organisation of Economic Co-operation and Development (OECD) member states. For instance, the objectives of Australia's *Telecommunications Act* of 1997 are: to promote the 'long term interests' of users (suggesting a concern with orderly development); to ensure that 'services of social importance' are accessible and supplied at standards that meet needs of the user community; and to promote a 'telecommunications industry that is efficient, competitive and responsive to the needs of the Australian community.'

The observations and questions set forth above provide a link between public policy objectives and the three generic intervention strategies, establishing a foundation for empirical research using interconnection as the basis for studying growth and change in large technical systems.

Three Facets of Wireless Enhanced 9-1-1

The case of Wireless E9-1-1 in North America presents an interesting window on the dynamics of change in the public information infrastructure but its relevance and value as a case study for mitigation-oriented policy research more generally emerges from a detailed assessment of the interactions, events, and outstanding issues that were evident in the design and deployment of the service in its wider relationship to the public information infrastructure. Given the importance of the undertaking for public safety and in light of wider policy objectives, what types of intervention strategies were employed, at what stage, and to what effect? As I have already noted, public policy objectives in current legislation suggest a possible role for each of the three generic intervention strategies: technology forcing, strategic niche management, and loci for reflexivity. What can we learn from this case that may assist in future decisions about using such strategies to shape growth and change in critical infrastructure?

It turned out that the deployment of Wireless E9-1-1 in Canada had a significant impact across the entire network infrastructure, affecting core and peripheral elements as well processes spanning all four levels in Arnbak's functional systems model. It also presented a fundamental public policy challenge in the emerging domain of mobile communications—a domain that hitherto had been largely characterized by regulatory forbearance in Canada—and demonstrated the complicated nature of technology forcing strategies in the face of technology convergence and efforts at balancing competition with public interest.

This initial facet of the Wireless E9-1-1 case is taken up in chapter four, where the difficult job of overseeing the development and deployment of technical standards is examined. In other respects, the development of Wireless E9-1-1 in Canada provides an exemplary illustration of the ability of interested parties to work together voluntarily through established forums; however, a persistent disagreement over problem formulations and design propositions imposed severe delays on the deployment of the service and eventually required regulatory intervention. This initial facet of the case provokes a consideration of how strategic niche management can work in conjunction with private commercial strategies to achieve public policy objectives, and this question is taken up in chapter five. The third facet of the Wireless E9-1-1 case examines the problems and delays in deploying the service while drawing out the inherent challenges of expanding stakeholder participation in consultative forums for setting public policy. As such challenges might be expected with intervention strategies based on loci for reflexivity, chapter six outlines the importance of establishing common frames of reference through appropriate forums for the sharing and exchange of views.

The purpose of the case study is to shed light on the processes and interventions by which network evolution occurs with respect to the management of critical infrastructure. I acknowledge at the outset that for some readers, Wireless E9-1-1 may seem an unusual case with which to consider the wider field of critical infrastructure. I would nevertheless contend that it is an innovation of interest for mitigation-oriented policy research insofar as it lends itself to a long-term program of risk reduction where, according to the Pressure and Release model, fundamental processes of growth and change in the Canadian telecom system are explicitly directed toward the objective of reducing unsafe conditions within society.

A Note Concerning Research Design

Data for the case study is drawn from primary-source document analysis, using an approach known as grounded-theory (Strauss and Corbin, 1990). My objective during the investigation was to produce a thick description of the events that surrounded the original demand articulation for Wireless E9-1-1, its introduction and subsequent development in Canada. Research began with CRTC Public Notice 2001-110, and I initiated the case study by registering with the CRTC as an interested party to the proceedings, thereby establishing myself as a participant-observer and obtaining directly all comments filed by other interested parties. Three rounds of comments were received from ten parties. The first round of comments (13 December 2001) was followed by reply comments (17 January 2002) and then further reply comments (28 January 2002). I reviewed in considerable depth all comments submitted by all parties, making extensive notes in two categories of interest in a process known in grounded-theory as 'inductive' or 'open' coding (Bernard, 2000, p. 444):

- Issues and arguments
- Key actors.

The category of issues and arguments refers to concerns raised by interested parties during the proceedings. Some issues were given greater weight than others, according to both the quantity and quality of effort given over to them. For instance, the matter of customer billing records being entered in the Automatic Location Identification (ALI) database produced the greatest debate in terms of word count, but it also brought the most forceful and provocative language between the interested parties.

Among the variety of key forces shaping the development of Wireless E9-1-1 in Canada, I tried to maintain a symmetrical perspective on both social and technical actors. Social actors were typically those delegates speaking on behalf of their organization and contributing to the proceedings. Technical actors in this case were those elements introduced or enrolled into the proceedings by social actors when articulating their demands, formulating problems, or asserting design propositions. Examples of technical actors include federal statutes and regulations (e.g., privacy legislation), specific regulations, technical standards, network and terminal equipment, professional practices, and so forth. I only considered these to be 'actors' in the proceedings if they were actively enrolled in a document that made a direct contribution to the proceedings in some form or another.

The vehicles of inscription on which non-human actors were established often served as the basis for subsequent rounds of investigation in my document analysis. For instance, where an interested party cited a previous CRTC decision or a published technical standard, I would note the context and obtain the associated documents. The process continued for several iterations until the point where the subsequent documents were not primarily concerned with Wireless E9-1-1 as their subject matter or where the influence on those documents extended beyond the influence of Canadian authorities. In some cases, I reached a 'trailhead' that represented a logical point to terminate a thread. For instance, a number of interested parties cited Canada's privacy legislation. Given that the subject of privacy legislation is not primarily E9-1-1, it seemed a logical point of termination for that particular thread of inquiry. In other cases, references to a document issued by the U.S. National Emergency Number Association (NENA) were common in the proceedings. In these instances I would terminate the document trail with those NENA-produced documents because they represent the edge of the Canadian PSTN boundary (as established in this chapter).

The telling of the case study is derived from deductive coding based on the generic intervention strategies of technology forcing, strategic niche management, and loci for intervention. Finer points of analysis were coded using the other two dimensions of the intervention matrix. This hybrid approach of inductive/deductive coding is suited to grounded-theory studies that incorporate a general thematic framework such as that presented by the intervention matrix (Bernard, 2000, p. 445).

In all, my case study examined every public document directly associated and cited in conjunction with the development of Wireless E9-1-1 in Canada, an extensive range of technical and policy documents produced by U.S. public safety organizations and industry groups, numerous FCC documents and related American legislation, as well as a score of secondary sources on telecommunications systems engineering (in order to understand terms and concepts used in the proceedings). Individuals from several organizations involved in the proceedings were contacted by email or telephone during the course of my study to clarify questions and in some cases to obtain documents that were not otherwise available.

Summary

This chapter began by first addressing the challenge of undertaking a technology assessment of an evolving, large technical system. Second, the chapter confronted the problem of establishing an operational basis for applying the intervention matrix to empirical study. With respect to these matters, I discussed some of the key features of technical networks and large technical systems, and considered the problem of boundary conditions as it relates to communications infrastructure. My objective in this chapter has been to expand the analytic framework along a third dimension of interconnection space, while introducing a number of current issues related to growth and change in the public information infrastructure.

Interconnection space is an important extension to the intervention matrix, as it draws out a set of specific issues in the areas of technical standards, interconnection arrangements, and expanded stakeholder participation. Together, the modified intervention matrix and related interconnection issues were used to inform an empirical study in public safety telecommunications, the results of which are reported thematically in the following chapters based on 'three facets' evident in the Wireless E9-1-1 case.

Chapter 4

The Standardization Effort

By the late 1990s the widespread diffusion of mobile telephones in North America began to pose a threat to public safety. Aside from the perceived health risks of increased exposure to radio emissions and concerns over road safety, a lesser-known threat has come from the unexpected impact of mobile phones on emergency services. Recent figures show that the majority of calls placed to emergency services will soon come from mobile phones, presenting a major challenge for the provision of public safety telecommunications (National Emergency Number Association, 2003). A pressing concern for emergency services operators when dealing with such calls is the ability to identify the geographic location of the person in distress because callers may not be able to report their specific location. As one might imagine, the difficulty in determining the geographic location of a mobile phone user can cause life-threatening delays for emergency dispatch and response. In other cases, 'emergency' calls made from mobile phones are not emergencies at all, but result from misdials that occur when mobile phones are wedged into handbags or purses. In these so-called 'trouble not known' cases, public safety agencies are often obliged to initiate a follow-up investigation which is made difficult or impossible when caller-ID or location information is unavailable.

The wireless enhanced emergency service ('Wireless E9-1-1') being introduced into the public telephone network across North America is a new service innovation that provides the geographic location and caller-ID information of a mobile phone subscriber in contact with an emergency operator. It promises to assist emergency services and to dramatically enhance public safety through improved dispatch and response. Its development also sheds light on the intervention strategy of *technology forcing*, raising important considerations on some of the finer points of this intervention strategy as it might apply more generally to the management of critical infrastructure.

The Genesis of Wireless E9-1-1

Over the course of some thirty-five years, public safety groups in the United States developed a rather advanced system to provide the public with the capability to request emergency services by dialling the digits '9-1-1' (National Emergency Number Association, 2002a). The original 9-1-1 concept was to provide a simple voice connection between the caller and an operator designated to handle emergency calls and dispatch appropriate agencies as required. In the early days of the service, this operator was often the local fire or police department but in many large centres today this function has evolved into a specialist role provided by a separate municipal or regional agency. With the subsequent development of more sophisticated telecommunications services an enhanced 'E9-1-1' system has become available, in which the voice connection is augmented with caller-ID and street address information associated with the telephone number of the handset used to make the call (National Emergency Number Association, 2002b). With this additional information provided in real-time to the operator, emergency calls can be quickly directed to appropriate jurisdictions for dispatch and in the event a caller is unable to provide their location verbally to the operator, emergency personnel can be given precise directions on the point of origin of the call. In the case of trouble not known calls, emergency operators are able to use caller-ID information to support follow-up investigations. This combination of Automatic Number Identification and Automatic Location Identification—or ANI/ALI (pronounced 'Annie-Alley')—functionality has since become the benchmark feature of E9-1-1 across North America.

With the rapid uptake of mobile phones beginning in the mid-1990s, the well-established E9-1-1 system quickly became fragmented between wireline calls with the ANI/ALI capability and those calls placed from mobile phones, which at the time provided no enhanced functionality whatsoever. This situation arose in the telecommunications infrastructure largely because the original E9-1-1 service had been conceived of within an entirely wireline environment in which virtually all customers had a telephone number that was more or less permanently associated with a specific street address, either at home or at a place of business. As a result, it was relatively easy to create a stable database of telephone numbers linked to municipal addresses. Mobile phones undermine this initial E9-1-1 concept, because they are not necessarily associated with a fixed physical location. Moreover, the commonly employed trunk-side routing arrangements between wireless carriers, incumbent carriers, and the public safety answering points have until only recently made it technically impossible to provide the enhanced functionality for mobile phones.

The spectacular popularity of the mobile telephone is an instance that illustrates the disruptive potential of technological innovation when it comes into contact with existing infrastructure and service design. What is perhaps most interesting about the case is the irony of it. For instance, mobile phones represent a leading edge telecommunications service using digital technology and advanced network capabilities to provide many of their value-added functions; yet, much of the initial core network arrangements put in place were simply not capable of providing even simple caller-ID functionality between mobile phone customers and emergency operations centres. Here we find that situation described by Thomas Hughes' as a 'reverse salient' in which innovations at the edges and upper layers of the network result in disruptive anomalies appearing at the core and lower layers of interconnection space. The cellular mobile phone network is an innovation added, as it were, to the edges of the existing telecommunications infrastructure and initially perceived as something of a complementary technology to the mainstream wireline service. The disruptive aspect of mobile phone technology stems from its widespread diffusion and subsequent impact on public safety telecommunications. A similar case that is now entering the regulatory discourse is Voice over Internet Protocol (VoIP) telephony, which also presents a number of problematic challenges to the conventional design of telecommunications services, including public safety.

In the United States, the Federal Communications Commission (FCC) was determined to solve the reverse salient problem known as 'Wireless E9-1-1' by setting out a two-phase strategy to force wireless carriers to develop and deploy the ANI/ALI capability. The strategy originated with the FCC's *Notice of Proposed Rule-Making* under Docket 94-102, issued in 1994. The primary intent of this Notice was to launch a series of related initiatives to ensure that the public information infrastructure in the United States would evolve along lines that would correspond with the emerging challenges to public safety telecommunications created by the diffusion of mobile phones.

Table 4.1 FCC Wireless E9-1-1 requirements

FCC Phase	Requirement
0	Transmit all mobile 9-1-1 calls to a PSAP
1	Transmit ANI and cell-site location with all 9-1-1 calls
2	Transmit ANI and lat./long. details of caller's location

During the first phase of the strategy U.S. wireless carriers were to develop a system that would provide local public safety answering points (PSAPs) with the mobile caller-ID number (ANI) plus low-resolution location information (ALI) in the form of the originating cell-site address. The more rigorous second phase requirements demand that wireless carriers implement a system that provides ANI plus high-resolution location information in the form of near real-time latitude/longitude coordinates of the originating mobile phone at the time the call is placed to 9-1-1 (United States Federal Communications Commission, 2001). High-resolution is defined according to the type of system that a carrier chooses to implement. For handset-based solutions, the FCC initially set an accuracy/reliability requirement of 50 metres for 67 per cent of calls and 150 metres for 95 per cent of calls. For network-based solutions, the FCC requires an accuracy of 100 metres for 67 per cent of calls and 300 metres for 95 per cent of calls. A typical handset-based solution uses a GPS (Geographic Positioning System) that involves placing a small receiver in the mobile phone to enable it to report its physical location using satellite-based radio signals. Network-based solutions, by contrast, use triangulation techniques calculated at cellular base stations or control points in the wireless network. One advantage of the network-based solution is that mobile phones need not be modified with GPS or other add-on features in order to be located in geographical space.[1] Phase One deployment was originally set for April 1998 and Phase Two for October 2001. Table 4.1 summarizes the FCC's Wireless E9-1-1 requirements.

For many in the U.S. wireless industry, Docket 94-102 inaugurated what was believed would be a rush to deploy other forms of commercial location-based services from the capabilities that Wireless E9-1-1 would bring to the mobile telecommunications sector. Numerous third-party vendors providers appeared on the horizon with applications developed in accordance with the FCC mandate, and in 2000 the Wireless Location Industry Association was formed as an advocacy body to promote location based services (Wireless Location Industry Association, 2001). Speculation in the trade press and much of the popular media at the time also reflected an optimistic outlook with respect to the future of mobile positioning and location-based services. In reality, however, progress has turned out to be slower than anticipated as many carriers filed for waivers on the Phase Two deadline and as of early 2004, almost three years after the original FCC target, most counties in the United States have yet to implement Phase Two deployments (National Emergency Number Association, 2004; United States Federal Communications Commission, 2003).

Technology Forcing and Wireless E9-1-1

The FCC mandate for Wireless E9-1-1 is a demand-side intervention strategy based on technology forcing and intended to steer developments in the public information infrastructure along a defined trajectory in order to accomplish wider public policy objectives. As introduced in chapter two, technology forcing is one of three generic strategies identified within Constructive Technology Assessment. It relies heavily on a vehicle of inscription to stipulate desired impacts and influence technology development accordingly. As we shall see further, the U.S. experience with technology forcing seems to be an exceptional approach to solving the Wireless E9-1-1 problem when compared with the strategies adopted in other jurisdictions.

Establishing a Vehicle of Inscription

A central concern for policymakers who have adopted a strategy of technology forcing is the question of determining the extent to which intervention is necessary to achieve public policy objectives. In the case of the FCC Wireless E9-1-1 mandate, the intervention must establish a carefully determined balance between the exigencies of the commercial telecom world and the objectives of enhanced public safety. Under such conditions, what instruments might a regulator or other institution employ to influence the evolutionary trajectory to ensure that public policy objectives are achieved while maintaining enough distance to ensure that the

---

[1] The website of the Alabama chapter of the National Emergency Number Association (NENA) offers a good explanation of these solutions: http://www.al911.org/

most appropriate solutions are developed and deployed? Put another way, what kind of *vehicle of inscription* is the most appropriate means to encourage an alignment of stakeholders around a stated objective while encouraging innovation through competition?

Under the FCC mandate, the primary vehicle of inscription has been a regulatory requirement that imposes deadlines and performance standards on wireless service providers. Actualization of those requirements has been largely hands-off, which on the face of it seems to be an appropriate strategy, given the argument that commercial service providers know how to best manage the technical details of their business. Moreover, within the current climate of regulatory reform, this is probably a necessary strategy. But has it been sufficient to achieve the public policy objective? Certainly the FCC mandate has encouraged a great deal of investment and activity in the technical development of location-based services for public safety. The very fact that it is now three years beyond the original deadline set for Phase Two deployment and so little deployment is evident across the United States would imply, however, that this intervention strategy has been far from sufficient.

Complementarity Oversight

The actualization of Wireless E9-1-1 involves a series of modifications to a large technical system, which suggests that interconnection is a primary consideration and as such, the nuances and trends associated with Arnbak's functional systems model ought to be considered closely. Furthermore, given the public interest at stake in the matter, there is also the question of selective regulatory intervention to ensure that stakeholder participation is adequate across the domain of interconnection space. As identified in the previous chapter, a fundamental basis for growth and change in large technical systems is the interconnection of services at multiple functional layers. In the segment of the design nexus, standardization is required to ensure interoperability of network elements and applications.

One facet of the Wireless E9-1-1 case reflects the strategic importance of the standardization process and more specifically, the question of whether or not a hands-off approach to the development of regulatory intervention is a sufficient public policy approach to the management of critical infrastructure. The FCC mandate provides an informative starting point because while it may have provided the incentive to develop the technical capabilities for delivering ANI/ALI from mobile wireless devices, it is not clear that is has succeeded in meeting the overall objective of making the service available to mobile phone customers. Details of the Canadian experience with Wireless E9-1-1 make this 'deployment gap' apparent and clearly illustrates the variability of concerns related to standardization across each of the layers of interconnection space. Observations of this kind are important for the more general question they raise about the extent to which the dynamic pressures associated with market liberalization and regulatory reform create a reluctance in regulatory agencies to become involved in the standardization process and to what extent this might perpetuate or lead to the formation of unsafe conditions. In other words, to what extent are the social roots of risk and vulnerability traceable to the technical standardization process? I hope to gain some purchase on this question by first sifting through some of the important dynamics and inherent tensions in the standardization process in order to put the question of regulatory intervention into wider perspective.

A Keystone Standardization Initiative

Richard Hawkins has coined the term 'keystone standardization initiative' to describe an undertaking that will 'have implications for those broad areas of technological development that relate directly to specific policy goals for [a] national telecom system' (Hawkins, 1997, p. 206). The significance of the term is in the attention it draws to those undertakings for which regulators need to be especially vigilant, in terms of monitoring developments and making interventions. Wireless E9-1-1 is a keystone standardization initiative insofar as it builds upon and establishes a set of technical and operational standards with significant implications for the future of public safety telecommunications and it is in this manner that it is related directly to public interest objectives.

In their most fundamental role, technical standards provide the foundation of interoperability for telecommunications systems and are therefore important prerequisites in the early development and deployment of many new services. This is certainly the case for Wireless E9-1-1 in the United States, where a major series of standards initiatives has been launched in the wake of the FCC mandate. Organizations such as the National Emergency Numbering Association (NENA), the Telecommunications Industry Association (TIA), and the American National Standards Institute (ANSI) have published volumes of recommendations and technical standards intended to enable functional requirements of Wireless E9-1-1 and to provide interoperability with existing systems.

Standardization is not merely a technical concern, however, as it increasingly has become regarded as one among many strategic inflection points within a competitive supply market, where key players seek to influence the design of networked infrastructure to their advantage (Mansell, 1993, 1999). Attempts to influence network architecture often begin with the standardization process, as Hawkins discovered with his work on European digital wireless telephony, where he observed a close link 'between standards and the development of particular services and product lines,' showing how standards are often deliberately introduced and heavily promoted at the initial design stage of a product or service in order to influence its position in the marketplace. Furthermore, in this strategic positioning role, standardization often serves as 'the institutional mechanism through which factions in the telecoms industry mediate between their adversarial commercial relationships and their common technical needs' (Hawkins, 1995, p. 33). These two observations represent what we might call *strategic tendencies* in the development of technical standards for new services, and both are borne out in the development of Wireless E9-1-1.

The challenge for regulatory intervention within a competition oriented policy framework is to establish a fine balance between standards directives and voluntary efforts to ensure that standardization initiatives are compatible with public policy objectives as set out in legislation. Yet, Hawkins (1997, p. 200) claims that while telecom reform initiatives have led to an increase in the range and complexity of standardization efforts, the direct involvement of national regulatory agencies in standards development is now more limited in scope than ever before. This observation notwithstanding, Hawkins introduces three principles to guide selective regulatory intervention in keystone standardization initiatives. The first principle speaks to the public interest at the most basic operational level, where regulators are responsible for ensuring the provision of public network services by issuing broad directives in the form of regulations. These kinds of services are meant to be available to all potential users on a reasonably equitable basis. Public safety services such as 9-1-1 are deemed to be in the public interest in most countries, and within an environment of rapid technological change it is necessary that regulators engage with standardization processes in order to 'maintain contact with the evolving technical parameters of different kinds of network services at the operational level' (Hawkins, 1997, p. 205).

There is a challenge to this role, however, and it arises from the expanded range of considerations that now motivate firms to participate in standardization following liberalization in telecommunications and other infrastructure sectors. To understand this challenge, it is first necessary to recognize that there are at least five 'key types of standardization rationale' that are active within a competitive infrastructure sector today: variety reduction, harmonization, intelligence, design, and market positioning (Hawkins, 1997, p. 201). Whereas variety reduction and harmonization are largely seen as matters of cutting costs through technical interoperability, the latter three rationales are more closely aligned with strategic business activities intended to better position equipment and service suppliers in a highly dynamic marketplace. Industry players may seek to participate in standards development to gain valuable intelligence on trends and issues, to influence the technological design in the early stage of development, or to directly link their marketing, research and development activities with a standardization initiative in order to improve market positioning (Hawkins, 1997, p. 204).

The second principle of regulatory intervention in standardization speaks to fair play aspects of competition policy, as regulators must ensure that standards developed for and applied at key network bottlenecks do not erect barriers to market entry, or create technological path dependencies that favour certain forms of service development over others. Regulators must find a way to engage with standards-making processes that ensures a balance between shifting technical exigencies and regulatory symmetry. The case of Wireless E9-1-1 suggests that national regulatory agencies have a role in ensuring that incumbent carriers do not use technical standards or the standardization process to abuse their inherited and exclusive control over the established 9-1-1 platforms that serve as intermediary network interfaces between the wireless service providers and the public safety answering points.

A number of instruments for the technical regulation of standards are available to prevent abuse of essential facilities but these must be appropriate to the context in which they are applied and they can be classified into three primary methods: voluntary standards, 'virtually mandatory' standards that dominate a market, and mandatory standards set by a regulatory agency (Hawkins, 1997, p. 205). When considering these three instruments, regulatory agencies must have a balanced influence on certain aspects of network development, especially those regarding general policy objectives, without imposing undue constraints on the standardization process that might counteract incentives to innovate and improve efficiency gains (Hawkins, 1997, p. 206). At the centre of this balancing act is a longstanding tension between voluntary and mandatory instruments. Typically, voluntary instruments are preferred by industry under the rationale that the industry is both the most competent and most motivated to work toward optimal designs. Yet, under these conditions 'optimal designs' may be rather subjective and could very well lead to problematic path dependencies as the dominant stakeholders seek to preserve competitive advantage in the market through proprietary arrangements (Hawkins, 1997, p. 198). In addition, uncertainty in the standardization process can result in delays in development and can pose subsequent barriers to investment and innovation in network infrastructure. For policymakers and regulators alike this tension amounts to a need to straddle direct involvement with an arms-length supervisory role.. With the emerging complexity of communications and other forms of digital technology, however, any form of broad scale influence is likely to extend beyond national borders and reside with a small number of large international equipment manufacturers, vendors, and service providers (Hawkins, 1997, p. 204). This suggests that domestic influence over standardization is seriously constrained by trends in the international domain, especially those initiatives led by American interests. Such is the case with Wireless E9-1-1, where American organizations have set the research agenda and have created a set of 'virtually mandatory' standards available for application in Canada and in other countries.

The third principle of regulatory intervention speaks to the broad social and economic policy objectives of national governments and the concomitant need for regulators 'to support this policy structure and ... to contribute to its development' by selective engagement with standardization processes (Hawkins, 1997, p. 205). Although specific wording will vary from country to country, most will have some form of policy that recognizes the vital role of telecommunications in key critical infrastructure and in supporting the needs of the population. In Canada, for instance, the *Telecommunications Act* clearly states that a primary national policy objective is to ensure the 'orderly development' of a telecommunications system that 'respond[s] to the economic and social requirements of users of telecommunications services' (Canada Department of Justice, 2001). In the United States, a legislative connection between telecommunications and public safety is made directly within the *Wireless Safety and Personal Communications Act* (also referred to as the '9-1-1 Act').

A keystone standardization initiative represents a window of opportunity to coordinate these wider public interest objectives with the competition directives of a liberalized telecom sector:

> Standardization is becoming a crucial node through which the technical design of network services will be self-regulated with the industry. ... the standardization phase is the only open point of access at which users can inject their collective needs and perspectives into the process of network and service design such that they might actually influence its outcome. (Hawkins, 1995, p. 34)

The key point here is that the standardization process provides an opportunity to involve not only users but also any other party that might be interested in participating in the growth and change of critical infrastructure. Standardization, in other words, offers an important point of entry into the design nexus, a chance to expand the discourse of infrastructure development, and a locus within which to encourage experimentation and assessment, thereby uniting the sustainability principle of mitigation-oriented policy with the economic interests of commercial actors.

Despite this opportunity, there are numerous conceptual difficulties in sorting out the multifaceted commercial relationships that may exist between users and suppliers as a result of network unbundling, which in a modular service scenario could very well provide multiple points of interconnection to new entrants and third party service providers. Such a situation results in a complex supply chain of services and creates the problem of distinguishing between suppliers and customers in a competitive services market. Hawkins addresses this problem by making a distinction between *end-users* and *intermediate users*, based on the observation that intermediaries extend the functionality of the basic public network facilities, while end-users typically adopt one or more service profiles provided by a carrier and represent a point of termination. .

Drawing on this distinction among user groups, Hawkins then describes some of the major practical obstacles to participation in standardization initiatives (1995, p. 28). Among these, he observes that the community of users tends to be fragmented in their views depending on their position in the supply chain, whether as end-users or intermediates. According to his findings, intermediate users tend toward a focus on upstream standards at the network core, while end-users concern themselves with downstream standardization efforts such as the terms and conditions of network access. This fragmentation splits resources and creates impediments for coherent participation in a keystone standardization initiative. In the discourse of a technology project, Hawkins' observations suggest that users will formulate problems differently according to their perspective on the network, which in turn may lead to competing design propositions. This scenario is all the more likely given the tendency of standardization to remain an upstream, supplier-led process 'without much active reference to service requirements as perceived by the users' (Hawkins, 1995, p. 33).

In response to these considerations, Hawkins (1995, p. 35) presents two alternatives to facilitate closer involvement of the user community in standards making. First he suggests the notion of an 'honest broker' or third party external to both supplier and user communities. The second alternative is that of 'the filter,' which is a body set up by a standards institution to allow the user community to participate in standards projects at various stages of development. Selective intervention is the key to these alternatives and some kind of incentive mechanism should be considered to ensure the establishment of 'user-administered requirements studies as constituent parts of the standardization process' (Hawkins, 1995, p. 36). A supervisory role in this regard is appropriate to the public regulatory function and is especially important in the case of a keystone standardization initiative such as Wireless E9-1-1, where downstream implications may be long term or otherwise far reaching for the public interest.

Regulatory Intervention and Interconnection Space

The historical shift in policy regimes from pro-incumbent to pro-competition has implications for keystone standardization initiatives. Within the pro-incumbent regime, networks were tightly integrated into closed vertical structures with limited points of interconnection among similarly positioned full service carriers. Technical standards were developed by a relatively small group of participants benefiting from a proprietary and vertically integrated arrangement. Recent developments in regulatory reform and a shift to the pro-competition regime have permitted a gradual dissolution of closed vertical structures and resulted in a more complex arrangement of network elements as new entrants are permitted to compete against the incumbent in the provision of a growing range of services at multiple points of interconnection.

The unbundling of network and service elements to support competitive access requirements leads to a potentially complicated arrangements of existing systems and emerging technologies. In effect, this means that a keystone standardization initiative may embody a diverse range of legacy systems and new standards under development by different organizations, which is indeed the case with Wireless E9-1-1. Therefore, having some means of sorting through the complexity of such a situation is important for regulators, first, to assess the possible limitations of industry self-regulation and second, to determine selective intervention strategies when such efforts fail to achieve policy objectives in a timely manner.

One such means for sorting out the complexity of emerging electronic communications infrastructures is to apply Arnbak's functional systems model, introduced in the previous chapter. Following a layer-based metaphor, the functional systems model classifies electronic communications services into a set of distinct but interconnected strata. In some cases, the layer model is put forward as a replacement to the traditional silo model used in communications policy where regulatory matters are more or less divided into separate vertical stovepipes such as voice telephony, radio communications, and broadcasting (Sicker, 2002). Within a layer model these vertical silos are replaced by a series of horizontal, functionally distinct but interacting subsystems.

The primary functional system is that which provides physical transport between source and receiver(s), and includes wireline and wireless forms of transmission as well as basic physical interfaces. The second system is that of network services, which includes the provision of switching, routing and gateway services. The third system is that of value-added services that provide client-server access to information content. Finally, the fourth functional system is for information services where content is created and supplied, often as a market good. With voice telephony, the third and fourth layers are less apparent. In the case of Wireless E9-1-1, however, the automatic generation of caller-ID and location information (ANI/ALI) are functions located in the application and information services layers.

By mapping Wireless E9-1-1 with the layer model, a range of supplier/user relationships and potentially varying degrees of regulatory supervision at each layer come to light. For example, a combination of wireline and wireless infrastructure (layer one) must enable end-to-end connectivity between a customer's mobile phone and the operator's desktop telephone at a public safety answering point (PSAP). Network services (layer two) must enable the correct routing of a 9-1-1 dialled call from the wireless service provider to the 9-1-1 platform and from there on to the PSAP. Network routing in layers one and two of the Wireless E9-1-1 system is divided into two main points of interconnection. The first of these is at the interface between the wireless service provider outgoing trunk lines and the 9-1-1 platform, which in some contexts is likely to be operated by the incumbent carrier, although this need not be the case. The second point of interconnection occurs at the interface between the 9-1-1 platform and the PSAP telephone system.

The provision of caller-ID and location information (ANI/ALI) between the wireless service provider and the public safety answering point can be classified into valued-added services (layer three) and information services (layer four). The value-added service component refers first, to the interconnection of signaling systems and databases that must occur in order to populate the 9-1-1 platform with ANI/ALI data and second, to the display of this data on the PSAP telephone system. My claim here is that this is a value-added function distinct from the 'basic services' network routing of the voice connection that takes place in layer two. Finally, information services refer to the process and procedures by which location information is generated by the wireless service provider, associated with municipal addressing schemes eventually formatted for entry into the ALI database at the 9-1-1 switch and, ultimately, for display at the 9-1-1 operator's computer terminal.

As this mapping suggests, for all its seamless functionality, Wireless E9-1-1 requires a complex arrangement of organizations and technical standards to enable and support service. At layer one, a physical infrastructure operator is required to provide end-to-end connectivity. In some cases where large distances or organizational boundaries are crossed, several layer one operators may be involved in the physical delivery of the voice call through switches and signaling systems (layer two). The wireless carrier, or perhaps a third party service provider, must operate an ALI database or 9-1-1 platform (layer three) that offers ANI/ALI data service to a PSAP. Finally, an information services provider must supply the location data in a raw or customized form either to the ANI/ALI database located at the 9-1-1 platform or directly to the PSAP telephone system (layer four). All the functional subsystems must be interconnected in order to have a working Wireless E9-1-1 service.

As one might imagine, the job of policy and regulatory analysis under such conditions can be difficult, particularly if numerous stakeholders are involved at one or more layers of the functional model and more so if strategic alignments among players vary according to the layer under consideration. Furthermore, some of these functions may be tightly regulated while others remain relatively unregulated. In the case of mobile telephones, for instance, the FCC has almost entirely forborne from the upper layers while maintaining an active supervisory role in the lower two layers of physical transport and network services through its Wireless E9-1-1 mandate.

Arnbak's model assembles these four distinct layers together into a multidimensional interconnection space useful for assessing unbundled network architectures. The results of such analysis not only help to make sense of a number of difficulties in coordinating the interests of parties involved in the deployment of Wireless E9-1-1; the analysis also suggests important considerations for regulatory intervention. By applying Hawkins observations about standardization with the full intervention matrix introduced in the previous chapters, it is possible to generate a set of introductory questions with which to inform both the case study of Wireless E9-1-1 and the empirical analysis of other forms of critical infrastructure:

- What are the current spaces of legitimacy for 'keystone standardization initiatives' and related technical developments?
- What is the basis for stakeholder participation (the principle of congruency) in these spaces? Is there a basis for regulatory intervention/supervision in these spaces?
- What forms of structural impediments or institutional barriers might constrain wider stakeholder involvement in standardization? What forms of regulatory instruments or other means might be drawn upon to address these barriers?
- How could an intervention strategy based on loci for reflexivity contribute to current standardization processes, and how might an 'honest broker' or 'filter' fit within such a strategy?

Soft Forcing and Wireless E9-1-1: the Canadian Case

Whereas the FCC in the U.S. introduced regulatory measures to promulgate a wider strategy of technology forcing for public safety telecommunications, the approach taken in other jurisdictions such as Canada, the European Union, and Australia has so far been less direct, where softer technology forcing strategies appear to be favoured. In the Canadian case, for instance, the development of Wireless E9-1-1 has largely been taken up as an industry-led voluntary initiative, although provisions do exist within current competition policy to provide the regulatory framework for technology forcing, though one that is applicable only under certain specific conditions.

Despite important differences with the American experience, the Canadian case provides useful lessons for other jurisdictions with respect to regulatory intervention. Foremost among these is that technology forcing must be seen as a multi-layered strategy that can work effectively with other intervention strategies if regulators maintain an effective supervisory role in the standardization process and take appropriate steps to act on the public interest where necessary. The following section will map the Canadian case against the four layers interconnection space to illustrate the complex nature of technology forcing as it applies to the influence of growth and change in telecommunications infrastructure.

The (non)Regulation of 9-1-1 Service

From a regulatory standpoint, emergency (9-1-1) service received what to many must seem a counter-intuitive treatment in the Local Competition Framework (Telecom Decision 97-8) issued in May 1997, which opened local telephone service in Canada to the prospect of competition at the local loop (Canadian Radio-television and Telecommunications Commission, 1997). In paragraph 113 of this decision, the Canadian Radio-television and Telecommunications Commission (CRTC) determined that 9-1-1 is not an 'essential facility' but then established in paragraph 286 'that it is [nevertheless] in the public interest to require CLECs [competitive local exchange carriers] to provide 9-1-1 service.' According to paragraph 74 of the Decision, to be 'essential' a facility, function or service must fulfil all three of the following criteria: it must be monopoly controlled; competing local exchange carriers must require it as an input to provide services; and competing local exchange carriers must not be able to duplicate it economically or technically. At the proceedings leading up to Decision 97-8, several parties argued that 9-1-1 should be treated as an essential facility but the Commission did not accept these arguments in its final determination.

With regard to *regulated* local exchange carriers, however, the Local Competition Framework does specify a quality of service requirement for 9-1-1 service that includes enhanced (E9-1-1) functionality in some cases:

> ...With regard to 9-1-1 service, all [regulated] service providers must ensure, to the extent technically feasible, that the appropriate end-user information is provided to the Automatic Location Identification database to the same extent as that provided by the ILEC [incumbent local exchange carrier]. (par. 286)

In the most basic terms, paragraph 286 of Decision 97-8 simply means that new entrants into the regulated local exchange market—known in the industry as Competitive Local Exchange Carriers (CLECs)—must conform to standards for providing emergency (9-1-1) service based on those set by the Incumbent Local Exchange Carrier (ILEC) in the respective operating territory. Yet, until 2003 these requirements did not have any bearing upon most wireless service providers because the wireless service providers do not qualify as regulated carriers. If a wireless service provider chooses to apply for CLEC status, however, then among the obligations set forth by the framework for local competition is a requirement to provide 9-1-1 service comparable to the ILEC-established standard. While this obligation may seem reasonable in exchange for certain advantages provided with CLEC status,[2] a literal implementation of paragraph 286 is problematic from the point of view of a CLEC seeking to offer *mobile* wireless service because the ILEC standard is based on a fixed address that is associated with a telephone number. The ALI function is thus generated based on a subscriber residence or business address in the database. This is clearly not as helpful when dealing with a mobile phone, where there is no necessary association between the address on a billing record and a subscriber's physical location at the time a call is placed to 9-1-1. Paragraph 286 embodies what Lessig (1999) has termed 'latent ambiguity,' which eventually surfaced in September 2000, when the CRTC granted interim approval to two new entrants—Clearnet PCS Inc. (later acquired by TELUS Mobility) and Microcell Connexions Inc. (using the brand name Fido)—as Wireless Competitive Local Exchange Carriers (W-CLECs). Telecom Orders 2000-830 and 2000-831 required these new entrants to introduce Wireless E9-1-1 where available, despite the fact that technical trials were still underway at the time the Orders were issued. Further, this service enhancement was not available in Canada nor was it likely become available for some time in many areas outside of major urban centres.

During the interim period, and in a misguided effort to avoid a situation of regulatory asymmetry, the CRTC directed the wireless CLECs to conform to the ILEC-established standard by populating the ALI databases with their subscriber records—despite the evidence that had been put to the Commission regarding the questionable value (and cost) of such an undertaking for a mobile service. In effect, the Orders sought to resolve the latent ambiguity of paragraph 286 by imposing an *ad hoc* interim solution that sparked more regulatory proceedings and dragged out a fractious debate over technical standards that led to considerable uncertainty and extended delays in the deployment of Wireless E9-1-1 in Canada.

The interim solution presented something of a 'soft' technology forcing strategy that revealed, among other things, the difficulty in determining appropriate and effective regulatory intervention when public interest is at stake in otherwise forborne areas of competitive activity. To understand more clearly the various dimensions of this problem, it is helpful to consider the range of specific issues that came to light during this particular standardization process.

Mixed Results and Shifting Alignments of Interest

Overall, the Wireless E9-1-1 experience in Canada has been characterized by mixed and often contradictory results that correspond with Hawkins' observations

---

[2] CLEC status offers carriers certain benefits unavailable to Wireless carriers such as local number portability (LNP) and access to contribution funds intended to offset service delivery in high cost serving areas.

on strategic tendencies in standardization. In spite of these results, the telecom sector and the public safety agencies did appear, at least initially, to work well together through the Canadian Wireless Telecommunications Association (CWTA) to adopt a set of voluntary technical standards for Wireless E9-1-1 (Canadian Wireless Telecommunications Association, 2003a). Undertaken in a relatively proactive manner and open to most major stakeholder groups, this voluntary approach proved to be effective insofar as it led to the introduction of Phase One capabilities on a timescale similar to that in the United States.[3] In fact, on numerous occasions participants from the wireless industry and the public safety agencies strongly advocated this voluntary approach as the best course of action, referring to problems being experienced in the U.S. under the FCC mandatory approach. For instance, one Canadian service provider claimed in a submission to the regulator in 2001 that 'wireless E9-1-1 rollout in the U.S. has been a disappointment' insofar as it was available only to a minority of the population. The submission cited a number of root causes which were apparently outlined in reports filed with FCC in June 1999, including problems in cost recovery, lack of liability protection for wireless carriers, the sheer number of stakeholders, and the compounded difficulty derived from a lack of cooperative spirit among stakeholders. Given these difficulties, the submission concluded that the FCC mandate 'has proved ineffective and may have been counter-productive' (TELUS Communications Inc., 2001a).

On the other hand and despite a series of successful technical trials, progress on the actual deployment of the service in Canada has been very uneven, particularly among the country's four major wireless service providers and especially outside its major urban centres where deployment of basic 9-1-1 service has been excruciatingly slow. Particularly evident in this case has been Hawkins' observation that under certain conditions 'the standardization process can result in delays … and pose subsequent barriers to investment and innovation in network infrastructure,' as deployment in Canada has hinged on the willingness of various interested parties to settle on a number of standards at all four layers of interconnection space. In a number of instances, certain parties have either been slow to make commitments or have called into question demands made by others within the proceedings.

While relatively simple in conception, Wireless E9-1-1 represents a significant service innovation in the telecommunications network infrastructure to include all dimensions of interconnection space. Moreover, decisions made within each layer, such as choice of signalling system, may impose going-forward constraints for other layers. Table 4.2 provides a summary of major issues drawn from the Canadian case, many of which may be used to generalize for issues in other countries.

---

[3] In Canada, the first commercial tariff for Phase One Wireless E9-1-1 service was approved by the CRTC in February 2001 (Canadian Radio-television and Telecommunications Commission, 2001g).

Table 4.2 Wireless E9-1-1 and interconnection space

Functional Layer	Considerations	Issues
Information Services	Subscriber records; cell-site mapping (ESRD); street address standards (MSAG)	Disruption of business models; control over critical information; customer privacy rights
Value-added Services	ALI database	Terms and conditions of access; database design and alternatives
Network Services	Network access services (Strawman; E911 tariff); ILEC/PSAP contracts	Terms and conditions of interconnection, including liability; requirements for new equipment; stranded investments
Physical Transport	Wireless local loop; cellular network; mobile handsets; trunking options	Problem equipment; legacy platforms; regional differences in network equipment and design

At the physical transport layer, the wireless local loop combined with cellular network design and mobile handsets has challenged the basic assumptions that were designed into the original database for generating automatic location information (ALI). The fact that a mobile handset on a cellular network is capable of roaming from place to place has created a problem in locating callers that simply does not exist with wireline service. Trunking arrangements between wireless service providers and the ILECs were also a central concern at this lower layer in the Canadian case, as they were unable to deliver the extra data elements needed to provide ANI/ALI to the public safety agencies. At the network services layer, there was a requirement to establish standard inter-carrier working agreements to specify interconnection arrangements and signalling between wireless service providers and the incumbent-controlled 9-1-1 system, as well as a need to negotiate additional contracts specifying network services and administrative details between wireless service providers, incumbents, and public safety answering points. The 9-1-1 platform and existing ANI/ALI standards represents a value-added data service at layer three that is operated by the incumbents to the public safety answering points. This service forms a key bottleneck in the provision of Wireless E9-1-1 and as such establishes a number of *de facto* mandatory standards at the upper layer of interconnection. At this upper layer of information services, interested parties discussed concerns such as the need to obtain and verify customer records for the ALI database, the accepted methods for cell-site mapping, and the requirements for standardizing municipal addresses ('Master Street Address Guide' or MSAG).

From an overall perspective on the case, it is apparent that industry voluntary efforts were most successful at the lower layers of interconnection space, in the physical transport and network services layers, where previously established and widely accepted technical standards could provide a relatively uncontested choice of, for example, signalling options and network interface arrangements. Evidence from technical trials and minutes from meetings of the industry-directed CRTC Interconnection Steering Committee (CISC) indicate that most parties involved in the technical trials agreed to adopt these standards with little difficulty, possibly leading one to conclude that technical harmonization may be more easily addressed in industry-led forums than those matters residing the upper layers of interconnection space.

The matter is not quite that simple, however. Standardization at the lower layers of interconnection space also appeared in one instance to have an influence on the business strategy of Microcell Connexions Inc.—known in Canada under its brand name 'Fido'—to become a Wireless Competitive Local Exchange Carrier ('Wireless CLEC'). Following a series of delays and a questionable ruling by the regulator regarding its E9-1-1 obligations, Microcell filed an application with the Canadian Radio-television and Telecommunications Commission—in part to request that it force the incumbent carriers in each province across Canada to commit to certain technical standards either immediately or in their upgrade paths—as a way to gain some assurances that the capability for deploying Wireless E9-1-1 across the country would be put in place within a specified time frame (Microcell Telecommunications Inc., 2001b). This request was, among other things, a strategic call for technical standardization in the lower layers of interconnection space. In its bid to evolve its operations from an unregulated wireless service provider to a regulated CLEC, Microcell may have been looking to approach the CRTC with some guarantee that it would be able to deploy Wireless E9-1-1 on a national basis within a reasonable period of time. This would only be possible if standard upgrades at the lower layers of interconnection space in the form of switching and signalling capabilities were mandated for the incumbent carriers who in most cases control the bottleneck that exists between the 9-1-1 platform and the public safety agencies. Such an assurance may have been seen by Microcell as a potential bargaining chip to avoid certain interim obligations set out in previously issued and highly controversial regulatory orders that had required any carrier with ambitions to become a wireless CLEC to provide access to their subscriber records in lieu of ANI/ALI data in cases where they were operating in a territory without Wireless E9-1-1 capabilities.

Lower Layer Interconnection: Microcell's Part VII Application

On 13 March 2001, Microcell filed an application with the CRTC, seeking relief in seven Canadian Provinces from barriers to the deployment of Wireless E9-1-1. This application and request was filed under Part VII of the *CRTC Telecommunications Rules of Procedure*, which sets forth the process by which certain kinds of applications for relief can be made to the Commission.[4] Under the Rules of Procedure, a Part VII application must contain a number of elements, including: 'a clear and concise statement of the nature of the order or decision applied for' and the title and section of the statute under which it is made; 'a clear and concise statement of the facts upon which the applicant relies;' and 'proposed directions on procedure where the applicant requests a particular form of proceeding or seeks to vary or supplement these rules' (Canadian Radio-television and Telecommunications Commission, 2001e). In other words, a party files a Part VII application with the CRTC as a means of requesting that specified action be taken to provide relief on a specific matter. The application may be directed against another party and a public procedure is specified to provide for interventions, reply comments, and interrogatories from the Commission and others. Part VII applications are highly structured, principally written, public proceedings on specific matters of telecom policy or regulation.

Microcell filed the Part VII application to request that the CRTC 'mandated provision of Wireless Enhanced 9-1-1 network access services … take action to ensure that other Incumbent Local Exchange Carriers … follow the lead of TELUS Communications Inc. … in providing network access services that permit wireless carriers to enhance their contribution to public safety in Canada' (Microcell Telecommunications Inc., 2001b). The timing of the application fell a month after the CRTC gave interim approval of the first commercial wireless E9-1-1 tariff and well before an Ontario-based trial had commenced. In its bid to become a wireless CLEC, Microcell was struggling with the directives and requirements given to it by the CRTC under Decision 2000-831: first, to adopt incumbent Bell's trunk arrangements, and second, to begin populating the ALI databases with subscriber information where Wireless E9-1-1 was not available.

These decisions perplexed many in the wireless industry for their sudden and largely unfounded ascendance in the regulatory proceedings. Microcell had opposed the Decision's directives on the grounds that Bell's proposed interconnection arrangements were only an *ad hoc* solution. Further, existing arrangements did not present a wise use of resources that would be better invested in testing and deploying true Phase One service. The ALI directive was contested on the grounds that it was also a waste of time and money to implement, especially given the perceived irrelevance of subscriber records for *mobile* emergency calls.

---

[4] The *CRTC Telecommunications Rules of Procedure* specifies several types of applications that may be made to the Commission by regulated companies and other parties. The term 'Part VII' refers to a general section within the Rules of Procedure and describes the procedure for a specific type of application. Part I describes the overall framework for the Rules of Procedure, but other application types are included under separate 'Parts' including applications by regulated companies for approval of tariff pages (Part II), general rate increases (Part III), certain applications concerning agreements and limitations of liability (Part IV), capital stock issues (Part V). Some application types address complaints by a subscriber or potential subscriber of a regulated company (Part VI). Part VII applications are intended to capture 'other applications by any person' and thus may deal with a wide variety of possible issues and concerns from an equally wide variety of parties (Canadian Radio-television and Telecommunications Commission, 2001e).

From Microcell's point of view, Decision 2000-831 amounted to a costly effort to implement temporary solutions while awaiting deployment of Wireless E9-1-1 service in its operating territories outside Alberta and British Columbia. The Order specified that Bell's trunk arrangements and ALI obligations would persist until Wireless E9-1-1 became available in an operating territory, at which point Microcell would be obligated to make E9-1-1 available to its subscribers in that territory. Comments submitted by Microcell indicate no objection to the E9-1-1 obligation but oppose the temporary arrangements imposed in 2000-831.

Of note is that the Part VII application did not request that other wireless carriers be mandated to provide E9-1-1 service, only that the ILECs be mandated to make available the requisite interconnection arrangements ('network access services'). From a strategic perspective, the Part VII application appeared in part to be a bid by Microcell to avoid implementation of the temporary obligations of Order 2000-831 by requesting that the CRTC adopt a soft technology forcing strategy to mandate Wireless E9-1-1 network access services in Canada. Such a mandate would provide certain assurance that Microcell could draw upon to argue its case *against* the feasibility of deploying temporary solutions in return for a guarantee that it would, as a wireless CLEC, deploy Wireless E9-1-1 on a national basis within a relatively pre-determined timeframe.

Microcell's application first established its claim to legitimacy[5] and then went on to describe the success of an Alberta-based trial, highlighting the early efforts in Western Canada as evidence for the feasibility of Wireless E9-1-1 deployment in other parts of the country. It then listed the major incumbent carriers from which it sought relief and specified their current and future capabilities to offer Wireless E9-1-1 service network access service. Upon considering the application, the CRTC issued a set of interrogatories intended to gather further information on matters raised in Microcell's case. Respondents to these interrogatories included the ILECs, several wireless carriers, and a handful of independent telephone companies in Ontario and British Columbia. None of the public safety agencies offered comments on the Part VII application, with the notable exception of the Montreal Urban Community/Union des municipalités du Québec (MUC/UMQ). The MUC/UMQ had submitted comments in the first round of the proceeding to request that the Commission modify the definition of Wireless E9-1-1 to include the real-time transmission of subscriber record information in addition to the 10-digit ANI (Communauté urbaine de Montréal and Union des Municipalités du Québec, 2001). These comments were later withdrawn in their entirety after Microcell sharply criticized the MUC/UMQ for being 'impractical, unhelpful and particularly disappointing given the opportunity they have wasted to support the advancement of public safety that the Application presents ...' (Microcell Telecommunications Inc., 2001c, par. 52). It is curious that the other PSAP

---

[5] According to Microcell, the basis for its Part VII application stems from section 32(g) of the *Telecommunications Act*, which sets out the general powers of the CRTC with respect to rates, facilities and services. The specific wording of 32(g) is an open provision that permits the Commission to 'determine any matter and make any order relating to the rates, tariffs, or telecommunications services of Canadian carriers.'

representatives, who had been so vehement in their stance on improving wireless 9-1-1 call provisions, remained utterly silent on the Part VII application, as it represented an opportunity to lobby for mandated Wireless E9-1-1 across Canada.

As required by CRTC rules of procedure, Microcell sought specific relief in the Part VII application, and this can be categorized into two types of remedies. In order to support the stated request 'that the Commission take action to ensure that Wireless Enhanced 9-1-1 is deployed expeditiously and to the greatest extent across Canada,' Microcell proposed a number of directives to this end. Building on the success of the Alberta trial and its technical design, Microcell framed its proposed directives in terms of the technical capabilities of the ILECs in each of the provinces. The TELUS design for Wireless E9-1-1 was based on CCS7 signalling and upgrades to their Nortel DMS 9-1-1 switching platform. Where a provincial ILEC was currently operating with a similar platform—as was the case with Ontario, Prince Edward Island, New Brunswick, and Saskatchewan—the application proposed that these ILECs file tariffs and supporting agreements within 60 days for the provision of Wireless E9-1-1 via CCS7 interconnection 'on equivalent terms and conditions to those approved for TELUS.' In cases where a different platform was in use, the application proposed different form of relief depending on the particular circumstances.

In the case of the province of Nova Scotia, the ILEC was at the time operating a Rockwell 9-1-1 switch but intended to replace it with a more sophisticated Nortel switch in 'the near future,' according to Microcell. In the meantime, however, the ILEC was providing its mobile affiliate with simple enhanced service using Feature Group C (FGC) trunk-side interconnection.[6] Microcell proposed two directives in this case: first, 'to immediately provide non-affiliated wireless carriers with a legal agreement and detailed implementation supported with documentation for FGC trunk-side interconnection to its existing 9-1-1 platform;' and second, to 'ensure' that Phase One interconnection arrangements were 'included as an integral part of its replacement [Nortel] 9-1-1 platform' along with the required tariff and supporting agreement for commercial provision of the service to non-affiliated carriers.

In the province of Manitoba, the ILEC was at the time operating a CML-manufactured 9-1-1 platform that, according to Microcell, was capable of supporting Feature Group D (FGD) trunk-side interconnection. FGD is capable of transmitting the 20-digit string necessary for FCC Phase One equivalent service and therefore provides a technical alternative to CCS7 interconnection. The Part VII application proposed that the ILEC be required by the CRTC to file a tariff and

---

[6] FGC interconnection service is most likely based on CAMA (Central Automatic Message Accounting) trunks, which transmit an 8-digit number in the signalling portion of the call set-up procedure. CAMA trunking was originally developed for recording and reporting of telephone call information at tandem switches. For Wireless E9-1-1, it can be adapted to provide a single digit NPA code number plus a seven-digit ESRD suitable for *routing* of mobile calls to the correct PSAP. This functionality is what is meant by the term 'simple enhanced service.' It provides neither ANI functionality nor true ALI functionality because cell-site/sector data is not made available to the PSAP operator.

supporting agreement within 60 days for FGD-based Wireless E9-1-1 network access service in its operating territory.

The only exceptional instance concerned the ILEC serving the province of Newfoundland. At the time of the Part VII application, this ILEC offered only basic 9-1-1 service in its operating territory, thereby rendering any form of enhanced 9-1-1 service (wireline or wireless) unfeasible because there was no switching platform to support it. Nevertheless, Microcell requested that the CRTC 'direct NewTel to ensure that Wireless Enhanced 9-1-1 network access service is included as an integral part of any future proposal to provide enhanced 9-1-1 service in its operating territory.'

As one might expect, the ILECs were unanimously opposed to a mandated approach for Wireless E9-1-1 network access services. Comments from them in the first round of proceedings cited numerous reasons to deny the application on the basis of several positions. The first argument was that Wireless E9-1-1 was a far more technically complex undertaking than portrayed by Microcell—not only would there be significant impact on the PSAPs in the process, also, associated costing and tariff issues would need to be closely examined. Further, the U.S. FCC mandated approach to Wireless E9-1-1 had not been successful and Microcell's proposed relief was contrary to certain provisions in the Local Competition framework. In one instance, Bell, in a joint submission with three other ILECs, argued that Microcell was 'trivializing' the efforts required for Wireless E9-1-1 interconnection by failing to properly account for different levels of software, different 9-1-1 system architecture and different equipment at the various PSAPs across the country. Bell added that 'modifications to the ALI output to the PSAPs are subject to a mandatory six to ten month terminal to network interface disclosure requirement,' calling into question the sixty-day timeline proposed by Microcell. Furthermore, Bell referred to comments that Microcell had itself made in the CLEC application testifying to the complexity of implementing even a partial enhancement of Wireless E9-1-1 service.[7] The Bell comments also claimed that Microcell's proposed timeline 'greatly understates' the 'labour intensive process' required to map cell sites to emergency service zones and to create corresponding ALI database entries (Aliant Telecom Inc. *et al.*, 2001a, par. 14-16, 21-26).

Bell Mobility expressed the view that a mandated approach was unwise, given the poor results achieved in the United States under the FCC requirements. To bolster its position it noted that 'less than 15 [per cent] of the U.S. wireless market has seen deployment of Phase One functionality' and that 'the adoption of a mandate has not produced the desired result.' Perhaps in an effort to give the CRTC pause for concern, Bell Mobility further pointed out that the FCC had faced

---

[7] It is important to note that Microcell had earlier claimed that Bell's offer of trunk-side interconnection was 'greatly oversimplifying the effort to implement even a partial enhancement of wireless 9-1-1.' These comments were made in an effort to demonstrate to the CRTC that interim E9-1-1 solutions were not a feasible approach to fulfilling paragraph 286 of Decision 97-8.

an additional burden of dealing with requests for waivers and mediating disputes associated with delays in deploying Wireless E9-1-1(Bell Mobility, 2001).

Despite being the first to offer a Wireless E9-1-1 tariff in Canada, the incumbent TELUS also opposed the Part VII application based on experience from the United States that suggested a mandate would be counter-productive due to a lack of coordination among stakeholders and that a 'joint planning approach ... ensures that all interests are accounted for and all solutions are explored.' Moreover, according to TELUS, such an industry-led approach would 'also [ensure] that the resulting service and tariffs will be consistent with the capabilities of the participants' (TELUS Communications Inc., 2001a, par. 18). TELUS also raised the matter of 9-1-1 service within the local competition framework established in CRTC Decision 97-8, arguing that because 9-1-1 service is not deemed an essential service under the terms of the Decision, the CRTC would be forcing ILECs to provide a competitive service offering. Even more problematic, TELUS claimed, was that forcing the wireless carriers to use an ILEC Wireless E9-1-1 service 'would have the effect of remonopolizing [*sic*] 9-1-1 service since wireless carriers would have no other option than to subscribe to the ILEC's service' (TELUS Communications Inc., 2001a, par. 20-21). Of course this claim was made contrary to the reality that 9-1-1 remains a *de facto* monopoly service and that the membership of the Canadian Wireless Telecommunications Association (CWTA) rejected outright Clearnet's proposal to introduce a third party 9-1-1 platform operator. Furthermore, TELUS' comments overlooked the fact that Microcell did not request such direction in its Application to the CRTC, rather, the request was that the ILECs make the service available to wireless carriers.

Microcell responded carefully to each set of objections to its requests under the Part VII application, claiming the current tariff offering in Western Canada proved that ILECs' technical arguments were 'a cover for inaction' because 'the differences that exist between the TELUS 9-1-1 platforms in [Alberta and BC] are as great or greater than the differences that exist between any other two 9-1-1 platforms in Canada' (Microcell Telecommunications Inc., 2001c, par. 6). In response to comments about the labour involved in deploying Wireless E9-1-1, Microcell cautioned the CRTC against falling for the ILEC's 'new found sensitivity to the workload of the PSAPs and wireless carriers in matters of wireless E9-1-1.' Microcell reminded the regulator of the way in which Bell strongly advocated for FGC trunk-side routing arrangements in the wireless CLEC tariff notice proceedings (later adopted by the CRTC as a directive in the Orders 2000-830/831). Noteworthy for Microcell was its claim that the FGC design requires 'precisely the same ESRD mapping and data entry processes that the Respondent ILECs now find so burdensome for PSAPs and wireless carriers to undertake' (par. 16). With respect to the FCC mandate, Microcell replied that the 'fundamental failing' of the U.S. approach was that while it required [wireless carriers] to transmit E9-1-1 data elements, it offered no means of ensuring that intervening 9-1-1 service platforms would be capable of handling these data elements. In support of this claim, Microcell offered anecdotal evidence to suggest that there had been cases in the United States where incumbent local carriers had refused to provide data interfaces to their ALI databases for incoming wireless E9-1-1 data (par. 29). By contrast, Microcell's proposed directives were 'expressly designed to overcome the fundamental failing of the FCC's approach' by ensuring that key bottlenecks to Wireless E9-1-1 service (i.e., the ILECs' E9-1-1 platforms) would be made available to wireless service providers *in advance* of any FCC-like mandate.[8] Microcell also took the opportunity in its reply comments to highlight what was in its view the 'obstinate refusal' of the Nova Scotia ILEC to provide legal and technical support documentation for currently available FGC trunk-side interconnection to support simple enhanced Wireless E9-1-1 service in that province (par. 40).

Having reviewed the Part VII application, the first round of comments, and Microcell's reply comments, the CRTC then saw fit to issue a set of interrogatories in an effort to gather further information on the matter of technical complexity and related timeframes for deployment. Two interrogatories were issued to wireline carriers—both ILECs and independent telephone companies—requesting information on a timeline for implementation of Wireless E9-1-1 and local E9-1-1 capability (Canadian Radio-television and Telecommunications Commission, 2001a, 2001b). One interrogatory was issued to the wireless carriers, including Microcell, and requested information on labour and a forecasted timeline for implementation. More specifically, the CRTC wanted more information on three questions. How long would it take wireline carriers to implement Wireless E9-1-1 network access services in their respective territories? What was the current capability of local PSAPs to receive and process Wireless E9-1-1 data? How long would it take wireless carriers to implement E9-1-1 if the ILECs and PSAPs were so equipped?

Responses to the CRTC's interrogatories were hardly helpful in sorting out the details of the ILECs' claims from the previous round. In part because there was little direction from the CRTC in how they wished the timeframe to be reported, the responses from the carriers seemed tailored to confuse rather than clarify the matter. The responses do not provide the standardized means of comparison that a regulator would need in order to arrive at a general consensus on the nature of the requirements and their respective timeframes. The ILECs and some of the wireless carriers, for example, provided highly detailed lists of tasks, with widely varying time requirements for the same kinds of tasks, while some of the independent telephone companies appeared to be poorly informed about Wireless E9-1-1.

---

[8] This is a subtle but important point of distinction that pertains to the design of Wireless E9-1-1 interconnection arrangements. Microcell's position suggests that the FCC mandate addressed the edges, or periphery, of the telecommunications network infrastructure and ignored the embedded core elements needed to ensure end-to-end interconnectivity. In policy terms, such a decision may be a deliberate attempt to promote the growth of alternative networking arrangements for 9-1-1 service. Likewise, it may also have been an unfortunate oversight given the fact that, as Microcell also pointed out in its reply comments, the U.S. situation involves 'a near mind-boggling fragmentation' of 9-1-1 service platforms and associated arrangements (Microcell Telecommunications Inc., 2001c, par. 33).

A most dubious matter in the responses to the interrogatory were claims by Bell and Rogers Wireless that data on the technical capabilities of the PSAPs fell under section 39 of the *Telecommunications Act*, arguing that 'release of this information on the public record would allow existing and potential competitors to formulate more effective marketing strategies and to focus on specific market segments, thereby prejudicing [Bell's] competitive position and causing specific direct harm to [Bell]' (Aliant Telecom Inc. *et al.*, 2001b). Astonishingly, the CRTC accepted this claim *prima facie* and permitted these companies to remove from the public record any information pertaining to the technical capabilities of PSAPs in their operating territories. Without such information, other interested parties seeking to prepare reply comments on the proceeding were extremely limited in their ability to verify the claims made by these carriers, a concern raised by Microcell in its later reply comments:

> That certain ILECs have the audacity to claim that summary data related to display technologies chosen and paid for by municipalities for the general public benefit is somehow proprietary to the ILEC's, speaks volumes about the degree of openness with which these ILECs approach 9-1-1 related discussions. (Microcell Telecommunications Inc., 2001d, par. 24)

Final reply comments for Microcell's Part VII application were submitted in early June 2001. Among other things, the application highlights the fact that consensus among the wireless carriers and ILECs is tenuous, despite the apparent success of the technical trial in Alberta and the cooperation that was then evident in the planning for the Ontario trial. All parties, with the exception of Microcell, recommended the Part VII application be denied because, as Bell put it, 'the material filed in this proceeding by the vast majority of participants confirms that the current [industry-led] approach followed for the implementation of Wireless E9-1-1 is the most appropriate' and that Microcell's proposed mandate 'would be impractical' (Aliant Telecom Inc. *et al.*, 2001c, par. 3).

Clearly, the ILECs and wireless carriers wanted to avoid a mandate and continue their work on Wireless E9-1-1 solely within the comfort of an industry setting such as the CWTA. Microcell's application for Wireless CLEC status had, however, forced the regulator's hand by drawing paragraph 286 of Decision 97-8 into the proceedings. As events unfolded it became clear with the Part VII application that while industry stakeholders were willing to work together to develop and test the technical capability for Wireless E9-1-1, an alignment of interests was less evident on more specific issues related to standardization.

Upper Layer Interconnection: Subscriber Records

In the upper layers of interconnection space, with value-added and information services, various concerns about standardization were also not easily resolved in the proceedings. An extensive debate over what data should be used to populate the ALI database involved reference to standards related to software design, municipal addressing schemes, and even the customer activation process used by mobile phone retailers. Details of this debate are introduced in chapter six but for our current purpose, the matter illustrates several considerations raised by Hawkins on standardization, especially the influence of large equipment manufacturers and international bodies, the problem of sorting suppliers from users in the telecom sector, and the challenge of fragmented user communities. More generally, while standardization does seem, as Hawkins has claimed, to serve as a mechanism to mediate factions in the telecom industry, adversarial commercial relationships are not necessarily the same at each layer of interconnection space. In other words, strategic alliances at one layer appear to disintegrate and reconstitute themselves in a variety of configurations that depend on the functional layer in question.

This became evident in the debate that took place over a design for the ALI database, illustrating the problem of external influences on domestic standardization initiatives and the fragmentation these influences may create among user groups. For instance, during the latter stages of the Wireless E9-1-1 proceedings, following a series of largely successful technical trials, the Public Safety Answering Points (PSAPs) insisted with increasing vehemence that the ALI database be populated with two types of addresses: one for the real-time location of the mobile phone customer and another for the customer's home or business address. For their part, the wireless service providers countered with the argument that a home or business address is irrelevant to a call placed from a mobile phone and moreover, that the current ALI database in use at the 9-1-1 platform is designed to accommodate only one address field per record. On this matter, the PSAP representatives (with one notable exception) were clearly aligned in their position as end users of the ANI/ALI system. As suppliers of the ANI/ALI service, the wireless service providers and incumbent carriers were aligned in opposition to the PSAPs. This set of alliances, essentially confined to layer two concerns regarding the design of network access services, became increasingly hardened as the E9-1-1 development proceeded from trials to deployment.

When one wireless service provider challenged the PSAP representatives to produce documentation on standardization that would support counter-claims being made about the ALI database capabilities to support multiple address fields, they eventually produced standards by the Telecommunications Industry Association (TIA) and the National Emergency Numbering Association, both U.S.-based organizations (Alberta E9-1-1 Advisory Association, 2002; Ontario 9-1-1 Advisory Board and Alberta 9-1-1 Advisory Association, 2001a). Microcell and the other wireless service providers rejected the evidence outright on the grounds that it ignored the reality of the ALI system in current use in Canada, dismissing the proposed standards as 'blue sky' visions (Microcell Telecommunications Inc., 2001a). Later in the proceedings, the ALI database debate fragmented parties within the PSAP community itself when the British Columbia 9-1-1 Service Providers Association, having had operational experience with Wireless E9-1-1, withdrew its call for populating the ALI database with a second address field and aligned itself with the wireless service providers and incumbent carriers (BC 9-1-1 Service Providers Association, 2001b).

This event illustrates the influence of American organizations such as TIA and NEAN on the development of this keystone standardization initiative in Canada.

Both organizations are well-resourced and influential participants that have published extensive standards documents for suppliers and users of E9-1-1 service. While the standards produced by these organizations are necessary points of reference for Canadian carriers and PSAPs, it appears that their predominance in the field has also created false expectations outside the U.S., which has resulted in significant differences in problem formulations among interested parties in Canada and in the subsequent design propositions put forward for the provision of ALI data from wireless carriers. American technical standards are forged in the particular working relationships unique to the U.S. and are not easily transposed into Canada where interested parties have very different histories and relationships with one another. That being the case, it is important to note too that North American interconnection arrangements more generally require that certain standardization regimes are adopted on both sides of the border, placing constraints on the range of design propositions that can be feasibly adopted in Canada.

The fact that a Wireless E9-1-1 system is designed around two major points of interconnection is an important consideration in this debate because it has been at the heart of numerous problems between the PSAPS and the wireless carriers, especially in the case of Microcell. In the overall process of providing Wireless E9-1-1 service, the supplier/user relationship is complicated by the two-point design, with the 9-1-1 Service Provider (usually the incumbent carrier) serving as an intermediary actor between the wireless carriers and the PSAPs. Constraints imposed by this intermediary arrangement create additional complexity in the design of the value-added services at layer two. Despite the fact that 9-1-1servicedoes not meet the basic conditions required for mandatory unbundling—at least according to competition policy in Canada—the 9-1-1 tandem and its associated elements will very likely remain a monopoly controlled and under the discretionary control of the incumbent carriers in each of their operating territories in Canada. The possibility of a competitive provision of E9-1-1 service was quickly extinguished in Canada early in the proceedings when a new entrant wireless carrier put forward a proposal that would have introduced a third party provider of ANI/ALI—a proposal that was not taken seriously by most other interested parties and eventually forgotten (Alberta 9-1-1 Advisory Association, 1999a; Ontario E9-1-1 Wireless Trial Committee, 2000).

A related matter, which is yet to be fully resolved, concerns the collection and standardization of customer records. This concern resides in the layer one domain of information services or 'content' of the Wireless E9-1-1 service design. The representative for Microcell went to great lengths in submissions to the regulator under Public Notice PN 2001-110 to detail the requirements of obtaining and verifying subscriber records during customer activation. As it turns out, a key technical factor in this matter is the Master Street Address Guide (MSAG), which pre-validates civic addresses before they are accepted by an ALI database. Microcell's position on the matter was that providing verifiable subscriber records would require real-time access to the MSAG and that such a proposition was fraught with problems resulting from the inconsistent nature of MSAG coverage in parts of Canada and the potential range of civic addresses that a mobile subscriber might provide when seeking service. In fact, Microcell had previously requested that Bell Canada, the incumbent operator in Ontario and Quebec, modify certain inter-carrier working agreements to include province-wide access to the MSAG in Ontario or, alternatively, that the regulator rescind its previous order for all wireless CLECs to populate the ALI database with subscriber records where E9-1-1 was unavailable. In this case the incumbent Bell, that was clearly on the side of the Microcell in the ALI debate with the PSAPs, now refused to cooperate with Microcell's request on the basis that unrestricted access to MSAGs 'could assist outside suppliers in producing revenue reducing competitive products' (Bell Canada, 2001b).

This matter illustrates another shift in strategic alliances between interested parties because it involves incumbent-controlled municipal street address data, which is an information service that supports ANI/ALI provisioning.[9]   When implementing an E9-1-1 service, local municipalities are often required to undertake a costly process of standardizing civic addresses to conform to the parameters of an MSAG system. In most areas of Canada this system is under the exclusive control of the incumbent operating the E9-1-1 platform. Microcell's request for wide area access to the MSAG was perhaps the first challenge to the exclusive control of a resource with potential commercial value for location-based services. Following Bell Canada's refusal to provide MSAG access to other wireless carriers, the representative for the Ontario regional 9-1-1 system complained bitterly that

> It is the view of the [Ontario 9-1-1 Advisory Board] and [other PSAPs] that municipal SAG information is provided to Bell Canada for their input into the 9-1-1 database to meet the needs of all municipalities with whom they have agreements for [E9-1-1 service]. Municipalities entrust Bell Canada with this information solely for the purpose of timely and effective emergency response. The intent is not to give them the right to claim sole ownership and distribution rights. We do not believe it is within their purview to unilaterally decide the extent to which this addressing data should or should not be made available to Wireless CLECs. (Ontario 9-1-1 Advisory Board and Communaute urbaine de Montreal, 2001)

In this instance, the strategic alliance of interested parties again changed, this time in the domain of information services as the incumbent carriers found themselves in opposition to the PSAPs and wireless service providers. The Ontario 9-1-1 Advisory Board's comments on this issue may prove to foreshadow future debates as more advanced mobile tracking capabilities are deployed in Canada. The MSAG is a highly accurate and up-to-date means of validating municipal street address information that could be used in turn to support commercial location-based services. It is also costly to maintain and municipalities may seek to recover some of these costs by licensing access to third parties, perhaps leading to further realignments between wireless service providers, incumbent operators

---

[9] For background on this technical detail readers may consult documentation produced by the CRTC Interconnection Steering Committee's Emergency Services Working Group (CRTC Interconnection Steering Committee, 2001).

and other commercial interests. These shifting alignments are summarized in Table 4.3.

Table 4.3 Shifting alignments in Wireless E9-1-1

Interconnection Space	Alignments	Conflict
Information services	ILECs versus WSPs and PSAPs	Access to Master Street Address Guide (MSAG)
Value-added services	ILECs and WSPs versus PSAPs	ALI database design
Network services	ILECs versus WSPs and PSAPs	Upgrade paths and timelines in certain provinces
Physical transport	Parties largely in agreement	Choice of trunking and signalling (minor disputes)

Summary: Strategic Considerations

The layer model helps to reveal how strategic considerations among stakeholder groups may shift according to the various dimensions of interconnection space. With Wireless E9-1-1, the unified effort between incumbent carriers, wireless service providers, and PSAPs that initially produced successful technical trials at the lower layers of interconnection space dissolved when dealing with matters in the upper layers. This may be partly a case of dominant players seeking to control the standardization of new services by delaying or otherwise refusing to provide access to certain network elements or information sources. In other cases it may be due to conflicting expectations among interested parties, as seemed to be the case with the PSAPs and the ALI database.

Under such circumstances or the threat of them developing, some measure of guaranteed and timely regulatory intervention must take place in order to ensure that public interest obligations are taken into account, especially in the case of matters of public safety. The layer model also suggests, however, that the most problematic standardization efforts may be located in the domain of value-added or information services, where the terms and conditions for regulatory intervention are often less certain.

The limited success of industry voluntary efforts in the case of Wireless E9-1-1 suggests that it may be possible to establish lower level interconnection arrangements without significant regulatory intervention. In terms of upgrading network services, however, incumbent carriers may be unwilling to commit in a timely manner to the upgrading of physical assets, such as the network access switches necessary for Wireless E9-1-1 service. In such cases there may be a need for a soft technology forcing strategy to ensure that basic infrastructure will be put into place in a non-discriminatory manner.

With upper layer issues, as with those surrounding the design of the ALI database, some of the misunderstandings and enmity between parties in Canada could have been avoided had the regulator assigned a third-party broker to more actively liaise between interested parties and E9-1-1 standardization activities taking place in the United States with organizations such as the TIA and NENA. Such a broker could have vetted detailed technical information and more closely assessed technical arguments within the proceedings, thereby providing an intermediary perspective on disputed issues and reliable information to all parties including the regulator. An intermediary position under these circumstances might have also assumed an overarching responsibility to identify and encourage all parties to adopt verifiable best practices in conjunction with critical path dependency issues related to network upgrade decisions and transitioning to Phase Two deployment of Wireless E9-1-1 in Canada.

As it happened, the design discourse was often contradictory and deliberately confused, with interested parties using the ambiguity of existing legislation and ineffective interrogatory procedures to their perceived advantages. The changing alignments of stakeholders across the various layers of interconnection space only served to complicate matters further. The CRTC's failure to intervene in a timely manner and its earlier decision in the Local Competition framework to regard 9-1-1 as a 'non-essential' service set the stage for the confrontation between the widespread diffusion of mobile telephones and the need for public safety. These are the kinds of dynamic pressures that may lead to increased risk and vulnerability as critical infrastructures experience growth and change. In order to probe the root causes of this dilemma, the case will next turn to look more closely at the technical trials as a form of strategic niche management.

Chapter 5

Innovation and Experimentation

Until 2003, the development of Wireless E9-1-1 in Canada had not been prompted by a direct regulatory mandate as it had in the United States. Yet, with more than 10-million wireless subscribers in Canada (Canadian Wireless Telecommunications Association, 2003b), the public safety answering points (PSAPs) responsible for handling 9-1-1 calls face similar problems to those of American emergency dispatch operators. Even before the CRTC issued regulatory directives in 1993, however, the policy framework in Canada had a powerful influence in shaping the development of a Canadian version of Wireless E9-1-1. The interplay between this regulatory context and industry-led efforts at innovation and experimentation form the background for the second facet of the case study in public safety telecommunications.

Mobile Phones and the Canadian Regulator

Under the *Telecommunications Act* of 1993, telecommunications in Canada falls within the portfolio of Industry Canada, and the Canadian Radio-television and Telecommunications Commission (CRTC) regulates many of the activities in this area. One of the central aims of the *Act* is to support the transition from a monopoly regime to a liberalized and competition-oriented telecom sector in Canada, with a specific mandate 'to foster increased reliance on market forces for the provision of telecommunications services.' While this mandate has set the general tone for the regulation of the wireless sector in Canada, section 34 of the Act has directly influenced the CRTC's decisions regarding wireless service providers in Canada. Section 34(2) is often a touchstone for Commission decisions regarding wireless service because it establishes the terms and conditions of regulatory forbearance:

> Where the Commission finds as a question of fact that a telecommunications service or class of services provided by a Canadian carrier is or will be subject to competition sufficient to protect the interests of users, *the Commission shall make a determination to refrain*, to the extent that it considers appropriate, conditionally or unconditionally, *from the exercise of any power or the performance of any duty* under sections 24, 25, 27, 29, and 31 in relation to the service or class of services. [emphasis added]

In other words, the CRTC is required to forebear from applying certain sections of the Act to those telecommunications services deemed to be within competitive markets. These specific sections fall under Part III of the Act and deal with rates, facilities and services, so include conditions of service, rates and tariffs, and working agreements between carriers.

Section 34 of the Act has proven to be a centrepiece for the future of the wireless sector through Telecom Decision 96-14, 'Regulation of Mobile Wireless Telecommunications Services,' which has established what we might call the second generation regulatory framework for commercial mobile telephony in Canada. This framework has now largely superseded the first generation regulatory regime for analog mobile telephone services established in the 1980s (Canadian Radio-television and Telecommunications Commission, 1996a). Decision 96-14 followed on the heels of the appearance of digital cellular service (also known as 'PCS') in the Canadian market in 1995 and the corresponding public consultation (Public Notice 96-2) to determine the appropriate regulatory classification and treatment of mobile wireless telecommunications services. The CRTC issued its final determination in December 1996, citing section 34 of the Telecommunications Act and stating that it would forebear from regulating most wireless service providers with respect to conditions of service, tariff approval, rates and calculation methods, and inter-carrier working agreements. This Decision established a second-generation, digital mobile phone service in Canada that would be largely unregulated by the CRTC, as it was deemed to be 'subject to competition sufficient to protect the interests of users.'

It is important to note, however, that at the time Decision 96-14 was issued, local exchange telephone service in Canada operated as a regulated monopoly held by the incumbent landline carriers, so the Decision in effect created a situation of regulatory asymmetry whereby wireless service providers, while providing local exchange telephone service, were not required to meet many of the obligations established for the incumbent wireline local exchange carriers (ILECs). At the time of the Decision, mobile phones remained a relatively marginal service but they have since become direct competitors to traditional telephone service as a form of *wireless local loop*, and yet wireless service providers are still not required to meet many of the public interest obligations set by the CRTC for local telephone service. Therefore, the choice by most wireless service providers to provide customer access to emergency services (9-1-1) was a voluntary offering that varied widely in consistency and reliability across the country. It was not until some years later that wireless access to emergency services (9-1-1) were confronted with more stringent regulatory requirements. Nevertheless, the wireless industry embarked on the development and deployment of Wireless E9-1-1 in the wake of the FCC mandate in the U.S. Unlike the American case, however, much of the early standardization efforts in Canada were, as I introduced in the previous chapter, entirely industry-led voluntary efforts.

Innovation Opportunities and Network Design

Research undertaken by Robin Mansell into early efforts at network unbundling (Mansell, 1990, 1993) presents a political economic perspective on the interconnection issue and implies an important role for public policy in supporting strategic niche management. Two central principles guide this perspective. The first principle is that the evolution of a network infrastructure is a process that should remain open to wide discussion and debate:

> Insofar as the telecommunications networks and services of the future will underpin the organization and use of information, the coordination and management of production systems, and new ways of gaining competitive advantage, it seems vitally important that the full implications of alternative telecommunications development trajectories be explored. (Mansell, 1990, p. 501)

Mansell's second principle claims that policy and regulation have a role in supporting new entrants as active contributors to growth and change in network infrastructure:

> Today's communication networks and services enable increasing opportunities for innovation and experimentation by smaller firms and consumers. Experimental initiatives by smaller firms and by public organizations contribute to the total stock of knowledge and competence available within each country. In a market characterized by ... dominant player[s]... the independent initiatives of public organizations and smaller firms must be candidates for public financial support to encourage experimentation on a broad scale. (Mansell, 1999, p. 96)

The first principle is clearly derived from a constructivist view of technology development, in that it posits multiple possibilities for the design and development of infrastructure, emphasizing the importance of considering alternative trajectories for development. The second principle ascribes to the societal learning objectives of Constructive Technology Assessment, where experimentation is encouraged as part of a public policy strategy to enhance overall capability. However, Mansell's principles tend to emphasize intervention based on strategic niche management rather than technology forcing. The basic idea is simply to encourage experimental initiatives through the public support of new entrants and other interested parties, based on the argument that they serve as positive risk takers contributing to new markets and promoting a rich diversity of services in the future development of telecommunications.

In light of mitigation-oriented policy research, Mansell's views suggest that an important area of study is that pertaining to innovation in network management and value-added services. One approach for mitigation strategies is to stimulate experimentation and testing of new services to support, for instance, business continuity planning and emergency response operations. Given the long-term trends of network evolution toward unbundled services and the growth of value-added information services, public policymakers might even consider ways to encourage the participation of third party service providers or seek to lure new entrants into the field to promote innovation. For such a strategy to succeed, interested parties require interconnection to various elements of network infrastructure. To be effective, a program of strategic niche management may require selective regulatory intervention in order to ensure that vital network elements are made available on fair terms and conditions to all interested parties.

The Strategic Scenario of Network Evolution

Mansell's research has demonstrated the 'non-neutrality of technical design' (Mansell, 1993, p. 207), meaning that interconnection arrangements are not simply a matter of technical decisions but a distribution of control among suppliers in the market, each seeking an advantage in design, operation, and use of network elements and services (Mansell, 1999, p. 87). She characterizes this non-neutrality in terms of idealist versus strategic scenarios, the former being one in which full competition is expected to prevail and the latter as one 'in which an oligopolistic market structure emerges where a few dominant players vie for success in the market' (p. 85). Based on a series of case studies on the development of so-called 'intelligent network' initiatives in Europe and North America during the early period of telecom reform, Mansell has concluded that the strategic scenario is orchestrated through a variety of design parameters affecting growth and change in networked infrastructure. As a result, policymakers and regulators 'need to evaluate the impact on social and economic goals of the uneven distribution of network control among a few oligopolistic players in the market' (p. 92).

From the point of view of path dependency and perhaps especially within Noam's 'system of systems' paradigm, the ongoing presence of a strategic scenario suggests a need for regulators to assess interconnection arrangements to assure careful balance between tendencies toward centralized control versus decentralized control of network elements and interfaces. This translates into three related tasks for policy and/or regulatory initiatives: one, the constraint of market power where dominant players are seeking anti-competitive or exclusionary arrangements; two, the creation of incentives for new market entry to promote diversity and competition; and three, the ensured coordination among multiple actors in supply of complex information and communication services to meet social and economic objectives (Mansell, 1999, p. 92).

The value of an idealist versus strategic scenario comparison is that it establishes a critical counterpoint to a view on interconnection that sometimes pervades thinking on innovation and public policy—namely, the idealist view that regulation is anathema to innovation. Mansell (1999, p. 88) draws on the strategic scenario to identify interconnection as a political economic issue, noting that it is not simply a commercial or technical matter, but a 'major bottleneck' that presents strategic opportunities for both incumbents and new entrants alike to consolidate or extend control over network infrastructure. This suggests that certain network elements may never be open to full competition without regulatory intervention or that access to critical network elements may require regulatory supervision to ensure fair terms and conditions for all parties in the interest of meeting security standards.[1]

In particular, there are at least two major areas in which control can be exercised through the design of interconnection arrangements: one, in access to underlying infrastructure; and two, in access to customers. On the one hand, we can classify these as control issues at the core and at the periphery of a network. According to Mansell (1999, p. 88), attempts to gain strategic control over networks have gradually shifted to the periphery, in the form of 'conditional access systems,' as open systems standards have become increasingly accepted (and mandated) for core infrastructure. On the other hand, we might also classify these as control issues emerging in the value-added services and information layers of interconnection space. For instance, Mansell has observed that control over access to customer information is an important strategy that may confer advantages in the market. In some cases, this mode of control is directly related to proprietary information generated by transactions at both the core and periphery of a network through proprietary interconnection arrangements.

Set against Arnbak's functional systems model, and considering the importance that both Melody and Noam have placed on effective unbundling for the development of competitive telecom services, Mansell's analysis reveals a subtle distinction found in interconnection design: while regulatory reform may be moving toward core infrastructure, efforts at strategic positioning appear to be migrating to the upper layers of interconnection space, toward network elements supporting value-added and information services layers. In one sense, these upper layers remain a kind of last bastion where dominant players can still retain a measure of proprietary control over service design and deployment. One simple example of this upward movement is evident in the Canadian setting, where unbundling directives from the regulator have opened much of the network core up to competitive interconnection; yet, discretionary control continues in value-added areas such as voicemail and certain operator services.[2]

If the idealist scenario of perfect competition represents a baseline in which conditions of supply are closely correlated with demand. By contrast, the strategic scenario is one in which dominant suppliers and major customers effectively control key aspects of network design to ensure a strategic advantage in the market. Accordingly, the role of policymakers and regulators in a competitive environment is to limit this control where it inhibits public policy objectives and to oversee interconnection arrangements in order to promote new entrants and to encourage experimentation and innovation where demand remains speculative.

---

[1] For instance, opening critical network elements within a competitive setting may hinge on the coercive power of government to ensure that all parties will conform to certain security standards, as with the use of legislation to ensure the protection of personal information.

[2] Voicemail systems do not interconnect between service providers and provide a form of conditional access system. Regulatory directives do not mandate interconnection of such value-added services at the periphery of the network.

Identifying Efforts at Strategic Positioning

The strategic scenario contributes to the analysis of growth and change in critical infrastructure because it proposes that certain parties will resist access to network elements where these elements act as operational control points or provide customer access. As a result, analysts should expect that dominant parties might attempt to deny access to elements or otherwise attempt to influence the development of a new service to conform to their wider strategic objectives. This possibility suggests that where dominant parties call into question alternative problem formulations or when they reject design propositions, the question of motivation, beyond technical arguments, should be carefully considered. To aid in this consideration we can draw further from Mansell's research to set out a list of indicators that may give evidence of strategic positioning through manipulation of specific interconnection arrangements (Mansell, 1993, p. 209). Table 5.1 summarizes these indicators.

Table 5.1 Indicators of strategic positioning through network design

Design Parameter	Indicators
Network interface standards	Attempts to maintain proprietary interface standards, network management software, and some aspects of service applications.
Unbundled intelligence	Resistance to requests from new entrants to provide access to intelligent network elements (i.e., SS7 elements); non-transparent pricing strategies; lack of responsiveness to small user requirements.
Product differentiation	Superficial variations in equipment design; strong differentiation in certain submarkets where competition is strongest, where cross-subsidies can be introduced, or where costs of innovation are highest.
Service competition	Weak competition in maintenance, billing, use of network resources and service applications; cross-subsidization between strong and weak competitive services; disparities in network access for majority of users.
Network access	Closed and uneven geographically distributed network access; restrictions on the use of public network resources; increasingly difficult negotiations over terms and conditions of interconnection.
Network control	Disparities in degree of network control available to service suppliers and different types of users.

Evidence of efforts at strategic positioning in contested problem formulations or design propositions may be detectable with this list of potential indicators. Analysis of growth and change in critical infrastructure that is focused in part on these combined indicators contributes to mitigation-oriented policy research efforts by identifying current interconnection arrangements that may constrain future efforts at strategic niche management. In order to explore possible limitations by interconnection arrangements on future strategic niche management efforts, we can return to the case study on Wireless E9-1-1 to take a closer look at its technical development.

Wireless E9-1-1 Technical Development in Canada

While the FCC in the United States had initiated an ambitious effort in technology forcing to create a service environment that would support the wider objectives of the *911 Act*, north of the American border the Government of Canada had not adopted national emergency number legislation nor were there plans to do so in the foreseeable future. The initiative to develop and deploy Wireless E9-1-1 in Canada was thus left up to the wireless industry, which turned to the matter in late spring, 1997, under the auspices of the Canadian Wireless Telecommunications Association (CWTA). Early functionality of wireless E9-1-1 in Canada would be achieved through a coordinated effort by the wireless industry, the incumbent carriers, and the Public Safety Answering Point (PSAP) representatives, working together cooperatively to sponsor a series of technical trials.

On 17 June 1997, a roundtable was held at the Airport Hilton in Toronto to affirm the wireless industry's commitment to developing wireless E9-1-1 for Canada. Previous to this event, the CWTA had conducted its E9-1-1 proceedings through an internal committee, which was now to be opened to include a wider constituency of stakeholders. Participants at the roundtable included members of the wireless industry, representatives from public safety answering points, equipment vendors, the incumbent wireline carriers, Industry Canada, and the CRTC. Agenda items at this inaugural event included the U.S. FCC mandate, the status of the Canadian industry, cost and cost recovery issues, consumer awareness, and a proposal for a joint CWTA/PSAP working group to further the development of Wireless E9-1-1 (Canadian Wireless Telecommunications Association, Wireless E9-1-1 Working Group, 1997a).

From a technical standpoint, participants at the roundtable identified the problem of providing mobile caller-ID—known more formally as 'Automatic Number Identification', or ANI—as a necessary and preliminary task to achieving full scale Wireless E9-1-1 functionality. Cost recovery was also deemed a major concern. In addition, the parties noted that consumer education was a priority, and the public safety answering point (PSAP) representatives agreed to share costs with the CWTA on developing a consumer awareness program to inform the public about wireless emergency calls. Together the parties affirmed that a working group could be formed and thus began a search for co-chairs representing both the CWTA and the regional PSAP groups. For the time being, it was determined that the development of Wireless E9-1-1 in Canada would not be a mandated undertaking as it had been in the United States, but rather, it appeared that the wireless industry would work to establish congruency among stakeholders to identify, align, and inscribe their interests in the design of a Wireless E9-1-1 solution.

Following the encouraging results of the preliminary roundtable, the CWTA formally established in July 1997 the Wireless E9-1-1 Working Group at a meeting of interested parties. In addition to selecting a co-chair from each of the CWTA and PSAP groups, the meeting produced a mandate for the Working Group, which served to inscribe the congruent interest among wireless service providers, ILECs, PSAPs, and others:

> The CWTA/PSAP E9-1-1 Working Group will examine the migration of E9-1-1 service in a wireless environment to give it the *equivalent capabilities of wireline E9-1-1* in locations where a PSAP is capable of receiving the information. The working group will identify, evaluate and prepare options for the migration of wireless E9-1-1 service in accordance with the priorities the working group identifies. [emphasis added] (Canadian Wireless Telecommunications Association, Wireless E9-1-1 Working Group, 1997b)

In effect, the Wireless E9-1-1 Working Group and its mandate would grow to become a foundation for a program of strategic niche management, as it moved from identifying and evaluating options for Wireless E9-1-1 to conducting technical trials in the Canadian provinces of Alberta and Ontario, and to a lesser extent in Nova Scotia.

As described in chapter two, strategic niche management is an intervention strategy defined in the field of Constructive Technology Assessment as 'the orchestration of the development and introduction of new technologies through setting up a series of experimental settings (niches) in which actors learn about the design, user needs, cultural and political acceptability, and other aspects' (Schot and Rip, 1996, p. 261). While on the surface the objective of the CWTA's working group appeared to be less about learning and more about developing a working system, the learning component was significant and it closely resembled in operational terms a process of strategic niche management. For instance, testing and technical trials dominated much of the Working Group's meetings from April 1998 through to the last reported meeting in July 2001.

A primary task to be addressed in the Working Group was simply to articulate a basic design proposition for Wireless E9-1-1. In the United States, problem formulation was central to the FCC rulemaking as Wireless E9-1-1 performance requirements were specified by the regulator and served as elementary design parameters. Without a similar framework in Canada, the first step for the Working Group was to establish a working definition for Wireless E9-1-1. The FCC's most demanding requirements for Phase Two deployment went largely untouched in the meetings, but FCC Phase One requirements were adopted as the model for Canadian Wireless E9-1-1. By this point in time, near consensus had been achieved in the United States on the matter that Phase One Wireless E9-1-1 would consist of two elements delivered to a Public Safety Answering Point (PSAP): a ten-digit mobile caller-ID number and a *pseudo*-ANI (Automatic Number

Identification) specifying the cell-site from which a mobile telephone call had been placed (Canadian Wireless Telecommunications Association, Wireless E9-1-1 Working Group, 1997c). In March 1998, the Working Group adopted the term 'Emergency Services Routing Digit' (ESRD) to replace the pseudo-ANI designation (Canadian Wireless Telecommunications Association, Wireless E9-1-1 Working Group, 1998c).

Following the American lead, it was clear among most stakeholders early in the Working Group's meetings that the first challenge for Wireless E9-1-1 was simply that of upgrading 9-1-1 interconnection arrangements to permit the 10-digit mobile caller-ID from a wireless subscriber to reach the incumbent-operated 9-1-1 platform so that it could then be routed to the appropriate PSAP. The ANI problem was thus given highest priority despite the fact that a number of incumbent carriers expressed their opinion that a solution to this caller-ID challenge was 'light years away' even in view of the fact that a location positioning solution could be implemented 'immediately' (Canadian Wireless Telecommunications Association, Wireless E9-1-1 Working Group, 1998b). Here was the first hint of resistance to early design propositions by a number of dominant players, although by no means were all incumbents in agreement on this point.

The initial difficulty appeared to be situated in the fact that then-current E9-1-1 tandems supported only wireline customers—an example of the 'reverse salient' problem. The problem is that wireline telephones are permanently associated with a fixed address residing within a designated Number Plan Area (NPA), also referred to as an area code, so the record field in the 9-1-1 database designated for caller-ID (in this application known as 'ANI') is typically designed for the seven digits that represent the customer's local exchange and specific line. The area code need not be included in each record associated with that local calling area, as this is typically assumed with wireline telephones. In situations where a region is divided into more than one area code, the 9-1-1 database will use an eight-digit combination that includes a single digit (e.g., 1-4) to represent the area code. Mobile terminals, however, introduce roamers into the local number plan area, which has created the need for a 10-digit record field. For example, a visitor from Toronto carrying her mobile phone when she is roaming in Vancouver (NPA 604/778) will have a 'local' number that includes NPA 416. For a Vancouver emergency operator to correctly identify this caller he/she will require access to the 10-digit number that includes the 'NPA 416.' The legacy ILEC 9-1-1 platforms were not designed to take into account the problem of wireless roaming and the 10-digit call back number represented, and for some interested parties, this was a significant and central technical obstacle of problem formulation for Wireless E9-1-1.

The Basic Design Proposition

Pseudo-ANI, or the Emergency Services Routing Digit (ESRD) as it has come to be known, seemed at the outset of the Working Group's proceedings to be a far less complicated problem to solve than the problem of the mobile ANI. ESRD is also a 10-digit number but it is associated with a cellular base station rather than a mobile telephone. ESRD provides a crude method of mobile positioning by linking to the geographic cell-site location from which a mobile call has been placed. The ESRD is a very low resolution positioning solution that is primarily intended to route mobile calls to the appropriate municipality for subsequent dispatch rather than to provide the definitive location of a mobile caller.

An important technical question for design propositions based on ESRD was whether an ESRD could be passed to the ILEC-operated 9-1-1 platform from a wireless carrier's switch and subsequently linked to the ALI (automatic location identification) field in the current E9-1-1 database. This novel arrangement would be necessary for ESRD to trigger the delivery of a cell-site/sector address instead of a wireline subscriber address. In this case, another technical barrier and reverse salient appeared as there seemed to be the need for two 10-digit fields to represent a mobile caller—one for the caller-ID and another for the ESRD—as opposed to one field for a wireline caller. In the wireline scenario, a caller's seven-digit telephone number serves as both the call back number and the trigger for the ALI database (this works because a wireline telephone number is more or less permanently assigned to a single physical location). In the case of mobile calls, however, caller-ID (ANI) is dynamically associated with an ESRD depending on the physical location of the mobile phone at the time the call is placed. This link between ANI and ESRD is then used to send real-time cell site/sector data to the emergency call-taker at the PSAP who can then ascertain the location of the caller. The question for the wireless service providers, ILECs, and PSAPs was whether or not current 9-1-1 platforms and display equipment were capable of being modified to meet the requirement for two 10-digit fields. Would the equipment be ESRD-compliant?

Early in the Working Group's proceedings, the question of ESRD seemed to be resolved. Two major incumbent carriers, Bell and TELUS, claimed that ESRD could be adopted in their respective operating territories, and new entrant Microcell (along with Bell) had already conducted captive trials of ESRD in Quebec prior to the formation of the Working Group. While there were some concerns expressed among its members as to their full understanding of the ESRD concept, in November 1997 it was nonetheless agreed upon as a fundamental design proposition for Wireless E9-1-1, and 'the adoption of pANI [ESRD] would be a necessary first step to routing 10-digit wireless phone numbers to the appropriate PSAP using [incumbent] Owner Companies' 9-1-1 platforms' (Canadian Wireless Telecommunications Association, Wireless E9-1-1 Working Group, 1997c). The CWTA membership as well as the PSAP representatives in the Working Group expressed a common desire to adopt a design that would provide a timely solution similar to the FCC Phase One requirements. The ESRD solution was also confronted with a cost factor that was equally on the minds of the CWTA membership:

> [The] CWTA expressed concern with any proposal that would limit the speed at which a Canadian solution equivalent to the FCC's Phase I requirement can be adopted. CWTA stated that the effort required to implement the pANI [ESRD] option is such that *little additional effort would be required* to arrive at a FCC Phase I equivalent

solution. [emphasis added] (Canadian Wireless Telecommunications Association, Wireless E9-1-1 Working Group, 1998b)

The proposed ESRD solution met with consensus among the members of the Working Group but subsequently raised the issue of numbering assignments, as each cell-site/sector would require a unique identifier that conformed to ALI database requirements. At the 14 November 14, 2000 meeting of the Working Group, incumbents Bell and TELUS indicated that they intended to propose the use of '5-1-1' as a prefix for the ESRD. In other words, cell-site/sectors would be assigned four unique digits preceded by a standard series that would include the local area code and '5-1-1' (e.g., 604-511-1234). This design proposition required enrolling an altogether different working group in the development of Wireless E9-1-1, as such prefixes cannot be assigned arbitrarily because they must conform to the North American Numbering Plan and other regional arrangements. Within the North American Numbering Plan (NANP) that provides the framework for a continent-wide telephone numbering system, an industry standard for special services is provided by the N11 code. For example, publicly accessible N11 codes are assigned in the following manner:

- 211 Public Information and Referral Services
- 311 Unassigned
- 411 Local Directory Assistance
- 511 Reserved
- 611 Carrier Repair Service
- 711 Message Relay Service (MRS)
- 811 Carrier Business Office
- 9-1-1 Emergency.

Canadian carriers are permitted to use 411, 611, or 811, whereas other N11 codes are made available for third party services. Prior to the technical trials for Wireless E9-1-1 in Canada, the 511 code had been

> held in reserve in Canada for access to Message Relay Services (MRS) by hearing persons who wish to communicate with deaf persons. Presently access to MRS by the hearing person is provided by a 1-800 number. Access to MRS by the deaf is provided by a 711 number. Consequently, in Canada, only 211 and 311 are currently available for assignment. (Canadian Radio-television and Telecommunications Commission, 2001f)

Requests for allocation and use of the remaining N11 codes are made through the CRTC, and typically handled by the CISC's Canadian Steering Committee on Numbering (CSCN).

As it turned out, however, the proposal to use '5-1-1' for the Emergency Services Routing Digit (ESRD) did not actually conflict with the previously reserved 511 service code for Message Relay Service, because of an important distinction between the ESRD function and that of a public access service. An

ESRD is a 'non-dialable' number, which means that it is only used within the Wireless E9-1-1 data network architecture and should not be confused with the public service 511 access code. A letter issued by the CRTC to the Peel Regional Police in conjunction with the Ontario Wireless E9-1-1 trial clarifies this important distinction:

> As you may be aware the Commission has recently initiated a public proceeding to examine issues relating to the assignment of the remaining N-1-1 service codes. It is noted that service code 5-1-1 is currently reserved and may be assigned in the future. It is understood however that the assignment of 5-1-1 as a service code will not conflict in any way with the use of 5-1-1 as proposed by the field trial participants, as a pseudo NXX, as ESRDs are non-dialable and do not trigger any routing.
>
> … given that the ESRD would not trigger any network routing, and that the use of NXX 5-1-1 as part of the ESRD would not impact or preclude any future use of NXX 5-1-1 as a service access code, there is nothing to preclude the use of NPA—5-1-1—XXXX as ESRDs in Canada. (Canadian Radio-television and Telecommunications Commission, 2001c)

The Working Group's proposal to use 5-1-1 for the ESRD was then assigned a Task Identification Form (TIF 37) by the Steering Committee on Numbering (CSCN) in early 2001. By July of that year the CSCN co-chair of the task force responsible for numbering reported that guidelines were being prepared for its implementation (Canadian Wireless Telecommunications Association, Wireless E9-1-1 Working Group, 2000, 2001a, 2001c, 2001d).

Signalling Methods and Interconnection

Having reached relatively early consensus on the ESRD as an acceptable solution to the ALI problem, the Working Group's membership turned its attention to the problem of delivering the combined ANI/ALI package from the wireless carriers through the network infrastructure to the PSAPs. The basic design proposition had been approved and it was now time to consider interconnection options.

Interconnection requirements for E9-1-1 can be divided into two functional layers that operate in two distinct legs: one with a voice component and another with a data component. In a typical E9-1-1 system, a voice call is routed to the appropriate PSAP based on a data service using the ANI. ANI is delivered from a wireline end-office switch to a selective routing database located at a 9-1-1 tandem switch, forming the *first leg* of the E9-1-1 system architecture. The *second leg* connects the 9-1-1 tandem switch to the PSAP telephone system. The ANI provides a function in each leg of this configuration. In the first leg, it acts as a trigger for the routing of the wireline call to the appropriate regional PSAP office. In the second leg, the ANI provides the PSAP call-taker with a visual display of the caller's telephone number, which can also be used to query an external ALI database. The ALI query obtains civic address information pertaining to the physical location of the telephone from which the call was placed. This ALI data may be linked in turn to a mapping application that will present the call-taker with a cartographic rendering of the caller's location. Figure 5.1 provides a simplified schematic of this basic network architecture.

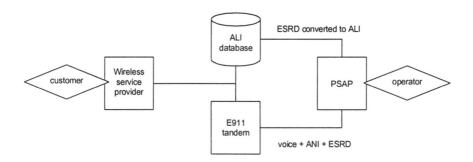

Figure 5.1 Simplified Wireless E9-1-1 call flow

A serious challenge for Wireless E9-1-1 interconnection stems from the fact that much of the currently installed PSAP equipment has been designed to work with a wireline-based ANI, often delivered in the second leg by means of call-path associated signalling (CAS). This 'in-band' interconnection method places a technical limit on the amount of data that can be provided with each call—often as few as eight digits. In the wireline environment this is all that is required to carry a single-digit area code signifier (usually a number from one to four) and a seven-digit telephone number. Much of the installed PSAP equipment is therefore designed to accommodate ANI based on this eight-digit string that is also used for querying the ALI database for location information.

When initial arrangements were made for wireless carriers to provide basic 9-1-1 service, this eight-digit number represented a trunk-group and *not the telephone number of any specific handset.* Moreover, this trunk-group, or 'pilot number' does not provide the easily recognized point of geographical reference that is needed to locate the caller and to route the call to the appropriate PSAP. In cases where a caller was unable to report his/her location, call-takers had to contact the security desk of the wireless carrier in order to request a manual search of the switch records for this information.

Signalling methods established initially for wireline E9-1-1 have thus had an impact on the development and deployment of Wireless E9-1-1. Each leg of the E9-1-1 system has been established with certain types of signalling technologies that constrain the kind of data that can be passed through the network. In simple terms, signalling can be in-band or out-of-band and typically falls into several basic types. In-band signalling uses multi-frequency (MF) tones sent ahead of the voice signal to provide basic information for call set-up and billing. This is also known as call-path associated signalling (CAS). Standard in-band signalling systems include a range of MF services sometimes referred to as 'Feature Groups.'

Each feature group has distinct signalling capabilities that may range from eight-digit (CAMA) to 10 or 20-digit enhanced MF (EMF). By contrast, out-of-band signalling—known also as non-call-path associated signalling (NCAS)—separates voice and data elements, with each corresponding to distinct network routing and interconnection arrangements. With ISDN and SS7-based designs, for instance, separate logical and sometimes physical pathways for voice and data are established between nodes in the E9-1-1 network (Lucent Technologies, 1999).[3] In effect, these methods differ only inasmuch as they present different means of delivering the ESRD and ANI data from a wireless carrier to a PSAP. The non-call-path associated design, however, introduces into the network some additional elements (and thus costs) between the physical transport and network access layers.

In the CWTA proceedings it became quickly evident that the wireless industry, PSAPs, and ILECs held different opinions on the best method of interconnection with respect to signalling options. Each of the methods would have a different impact on the costs for parties involved. In autumn of 1998, the Working Group began to address in earnest the issue of call-path versus non-call-path methods for Wireless E9-1-1 in Canada. The CWTA initially expressed support for a call-path signalling solution, whereas the PSAPs in Alberta and Ontario preferred a non-call-path solution, citing the cost of equipment upgrades as a major concern. Despite its position on the matter, the CWTA did recognize that a regional approach might need to be considered:

> The CWTA continues to be supportive of a CAS based solution for Wireless E9-1-1, however, the Association understands that in some areas of Canada a CAS based solution may not be feasible due to cost and/or operational impacts. Therefore, the CWTA is recommending that an evaluation of NCAS and hybrid solutions be conducted by the working group to address those regions where CAS cannot be supported. Along with this evaluation, the impact of regional solutions in comparison to the original objective of a common national approach must be identified. (Canadian Wireless Telecommunications Association, Wireless E9-1-1 Working Group, 1998d)

The incumbent carriers at least initially seemed to be divided on the matter, with some expressing the opinion that a hybrid solution would be the most cost effective and others refusing to make a commitment at the outset of discussions. One wireless carrier in the Working Group expressed concern over experience in the U.S. that seemed to indicate that the non-call-path solution was proving problematic from a technical standpoint, giving some credence to the CAS method. Minutes from the 28 September 1998 meeting of the Working Group also indicate that the CAS method was being considered for trials in Nova Scotia and Manitoba (Canadian Wireless Telecommunications Association, Wireless E9-1-1 Working Group, 1998d). Oddly enough, considering its potential implications on a number of fronts, the issue of signalling methods was never again raised in subsequent meetings. In fact, it was discussed in only three meetings of the Working Group between March and September 1998.

---

[3] ISDN (Integrated Digital Services Network) and SS7 are both digital systems that permit separation of voice from data signals.

Technical Trials

Central to the activities of the CWTA's Working Group was an attempt to foster a series of Wireless E9-1-1 technical trials in the Canadian provinces of Nova Scotia, Alberta and Ontario. The earliest mention of one of these is in the 10 November 1997 minutes in a brief discussion of a 'captive' trial undertaken by wireless carrier Microcell and the incumbent Bell in Quebec to test the feasibility of using *pseudo*ANI to route wireless emergency calls. It would not be until April of 1998, however, that the discussion of Wireless E9-1-1 trials would become a major pre-occupation for the Working Group. At the 28 April 2004 meeting, three incumbent carriers made presentations outlining possible approaches to trials. Several expressed a negative view on call-path associated signalling (CAS), claiming that it 'would not only degrade the integrity of the existing E9-1-1 network, but would also require a major network reworking in order to support a wireless E9-1-1 service that meets [FCC] Phase II requirements.' TELUS offered a presentation that introduced an NCAS approach for Alberta and yet another design proposition from the incumbent in Manitoba outlined a possible CAS approach for a trial in that province (Canadian Wireless Telecommunications Association, Wireless E9-1-1 Working Group, 1998a). Here it becomes apparent that a number of competing design propositions based on a range of different signalling methods were at play in the proceedings leading up to the launch of formal technical trials.

The operational objectives of the technical trials varied slightly according to the region in which they were conducted (see Table 5.2). For instance, Nova Scotia was among the earliest Canadian provinces to begin a Wireless E9-1-1 trial but focussed principally on a preliminary objective of simply routing wireless calls to appropriate PSAP regional offices through the use of a *pseudo*ANI-type arrangement. In this trial no attempt was made to actually delivery either *psuedo*ANI or caller-ID data to the PSAP office. On the other hand, the trial conducted in Alberta was far more ambitious in bringing together four wireless service providers and the Calgary PSAP to conduct a full-scale operation that included ESRD-based routing and full ANI/ALI display at the PSAP. Incumbent Bell in the province of Ontario followed suit with a similar trial in the Toronto area.[4]

The technical trials in Canada provided a foundation for strategic niche management insofar as they served as a prototypal design stage 'in which actors learn about the design, user needs, cultural and political acceptability, and other aspects' before committing to commercial deployment (Schot and Rip, 1996). Indeed, this was stated as a principal reason for undertaking the trials, which have

---

[4] An important reason for the difference between Nova Scotia and Alberta may be due to the wide variation in wireless coverage between these two regions at the time. Two wireless carriers served Nova Scotia, with limited digital cellular coverage confined to major urban centres. Alberta, on the other hand, was at the time well served by four wireless carriers, providing extensive analog and digital coverage across the province. As a result, it made more sense that the Alberta trial would provide a setting for more ambitious testing of Wireless E9-1-1.

proven to be informative from a variety of perspectives, a number of which I will address in detail using examples drawn largely from the Alberta Trial Report (Alberta E9-1-1 Advisory Association, 2000a).

Table 5.2 Major Wireless E9-1-1 technical trials in Canada

Province	Date Launched	Lead	Enhancement
Nova Scotia	Nov/98 to May/99 (6 months)	MT&T	ESRD (routing only)
Alberta	Oct/99 to Apr/00 (7 months)	TELUS	ESRD/ANI (to the PSAP)
Ontario	Jun/01 to Dec/01 (7 months)	Bell	ESRD/ANI (to the PSAP)

The Alberta trial officially commenced in October 1999 and was scheduled to continue until 31 January 2000, but was eventually extended to 30 April 2000. The extension was negotiated to give regional incumbent TELUS an opportunity to seek approval for a tariff filed with the regulator to offer commercial interconnection for Wireless E9-1-1 network access services. In effect, the extension was to allow the participants the option of a smooth transition from a trial to a commercial rollout of Wireless E9-1-1 by leveraging the existing architecture. In point of fact, however, the only wireless carrier to initially adopt the service would be Microcell, who by October of 2001 introduced Wireless E9-1-1 province-wide in Alberta and began testing it in the Greater Vancouver Regional District of British Columbia (Canadian Wireless Telecommunications Association, Wireless E9-1-1 Working Group, 2001b). With its CLEC application pending, Microcell had its reasons to move quickly to Wireless E9-1-1 but the other wireless carriers did not seem to share any sense of urgency following the trial's completion. Despite the apparent success of the trial and the subsequent commercial Wireless E9-1-1 tariff offering, the other competing wireless carriers in the province declined to offer their customers the enhanced service and—much to the dismay of the Alberta E9-1-1 Advisory Association (AEAA) who had spearheaded the Alberta trial—remained with only line-side interconnection for basic (voice only) 9-1-1 service for some time afterward.

Mounting the Alberta trial for Wireless E9-1-1 was foremost an undertaking in negotiation and cooperation among the participants. For what was to be a four-month long live trial (November 1998 to September 1999), it took some eleven months of planning to establish guidelines and most importantly, to finalize the wording on a Memorandum of Understanding (MOU). The author of the Alberta Trial Report echoes similar sentiments to those that were expressed in CWTA working group meetings, when he wrote

The single most difficult aspect of moving to the operational phase of the Trial was successfully negotiating a memorandum of understanding (MOU). ... The MOU and

Schedules ... were an exercise in compromise that took the efforts of all participants to reach successful and amicable solutions. (Alberta E9-1-1 Advisory Association, 2000a)

Furthermore, it is clear from related documentation that the leadership of the Alberta PSAP representative was paramount to the success of the trial as his position represented a neutral third party acting beyond the fray of the competitive telecommunications carriers. The MOU between the parties provided a legally binding vehicle of inscription necessary for the trial to take place, setting out in detail its exact terms and conditions, including the following provisions:

- Trial area (geographic boundaries)
- Trial architecture (interconnection)
- Trial location data record format and presentation
- Trial call answer arrangements (PSAP equipment)
- Responsibility for costs
- Trial plan (goals, considerations, and implementation)
- Limitations of liability.

The trial area was confined to a north portion of the City of Calgary, including the neighbouring rural areas, but used only one E9-1-1 tandem, thereby obviating the need for more complicated inter-tandem networking arrangements. Bell's later trial in Ontario, while similar in many respects to the Alberta trial, sought to include an inter-tandem component to test the additional functional requirements of an extended configuration. The Alberta trial involved only one regional area code, although the architecture successfully handled ANI from roaming handsets with other area codes. A major consideration at the outset of the trial, however, was the problem of unsubscribed mobile phones.

Although the wireless industry in Canada had in the interest of promoting public safety agreed to handle all 9-1-1 calls dialled from mobile phones whether or not the phones were associated with an active account, the obvious problem for a Wireless E9-1-1 service, however, was that unsubscribed handsets might transmit an invalid mobile identification number and it was not known at the time if it was possible to prevent these from being sent to the PSAP, as each of the wireless carriers' switches was expect to handle this situation differently. A similar problem had been raised in the CWTA Working Group's meetings with respect to the availability of so-called 'one button' phones that are, in effect, non-registered mobile handsets designed with the sole purpose of dialling 9-1-1 when a button is pressed. Such devices had been approved for sale in Canada as personal safety devices.

In dealing with this situation, one wireless carrier decided to route these exceptional calls over its existing line-side trunks so that the invalid number would not be transmitted to the PSAP. For their part, the other wireless carriers initially stated that their switches would not transmit a number for unsubscribed phones and therefore would handle such calls in no special manner. The trial report notes, however, 'it was discovered during the Trial that ... non-dialable [numbers] *were being transmitted* on calls from unsubscribed callers' contrary to what the wireless carriers had claimed would happen. Despite initial confusion at the PSAP, it happened that the non-dialable numbers were unique enough to be recognized by trained operators who quickly identified them with unsubscribed handsets [emphasis added] (Alberta E9-1-1 Advisory Association, 2000a, p. 9).

Technical details of the trial were discussed early in the planning stage, with a number of design propositions for signalling options. Based on previous experience in the Nova Scotia trial, Microcell put forth a plan to use conventional Feature Group C trunking that would permit quick implementation of a call-path associated signalling method. The incumbent TELUS also offered a method based on call-path associated signalling with Feature Group D trunking and 10-digit ESRD that would use existing interconnection and promised easy implementation 'with [a] low risk of stranded investment.' TELUS offered a second proposition, also apparently with a low risk of stranded investment, based on non-call-path associated signalling to deliver a 20-digit ESRD/ANI combination to the PSAP, noting that this design would require upgrading to their 9-1-1 tandem.

New entrant wireless carrier Clearnet PCS proposed a hybrid solution based on a non-call-path (ISDN-based) configuration to deliver ESRD to the PSAP. But the most interesting development in the Clearnet proposal was its later revision into a more radical design that provided for 'direct interconnection to PSAP's through a third party provider' that would bypass the incumbent-operated 9-1-1 tandem altogether. Other parties responded with a preference for a single interconnection methodology universally applied, thus ruling out the option of each wireless carrier choosing to adopt a unique method and effectively ruling out Clearnet's proposal for a competitive provider of an E9-1-1 platform (Alberta 9-1-1 Advisory Association, 1998, 1999a).

The TELUS proposal based on the hybrid arrangement was agreed to on 11 March 1999 and eventually implemented for the trial (Alberta 9-1-1 Advisory Association, 1999b). This decision by the trial participants may be attributed to its promise to deliver the full ANI/ALI package using existing interconnection arrangements, which would require little by way of additional investment for the wireless carriers. Current PSAP equipment in Calgary was also capable of accommodating the design. Much of the risk was taken by the incumbent TELUS who would have to upgrade their 9-1-1 tandem to deploy this method.

Reported results from the trial indicate that it was a resounding success. All the objectives and considerations were rated 'achieved' with the exception of cost recovery, which was to be dealt with later in the TELUS tariff application. The technical architecture proved viable and the Calgary PSAP found the enhancements improved its operations. Among the challenges and issues raised in the Report, several were anticipatory in nature, including the problem of updating ESRD records with cell-site coverage modifications, as well as anticipated difficulties with invalid numbers from unsubscribed handsets. Among the foreseeable problems, one that would come to haunt the later development of Wireless E9-1-1 in Canada was the seemingly innocuous problem of tracing the subscribers of disconnected calls:

With the provision of MIN, wireless carriers can expect to receive a larger number of requests for subscriber information related to 'trouble-not-known' calls. It is critical to ensure that 7x24 procedures between PSAPs and wireless carriers are firmly established. (Alberta E9-1-1 Advisory Association, 2000a)

This issue of tracing subscribers was later to become a source of contentious dispute among interested parties. The matter would eventually involve the CRTC and would require the regulator to reconsider its policy on the deployment of Wireless E9-1-1 and its regulatory forbearance in the wireless sector.

Deployment Delays

If indeed the Alberta Trial was such a success, then what of the deployment of Wireless E9-1-1 in the operating territory where it was undertaken? Why is it that only one among three major wireless carriers operating in this territory acted in a timely manner to introduce E9-1-1 service to its subscribers? These questions were in fact put to the competing wireless carriers at several meetings of the Working Group long after the trial had concluded. One wireless service provider persistently objected to a number of issues in the commercial tariff filing, noting problems with certain limitations of liability. Responding to a question at the 8 May 8 2001 meeting, TELUS Mobility replied 'that since the Microcell Press Release six weeks ago, [they] would not discuss their future plans for implementation in an open forum,' noting further that Microcell had 'moved 9-1-1 to a commercial matter vs. a non-commercial matter' (Canadian Wireless Telecommunications Association, Wireless E9-1-1 Working Group, 2001d).

The press release in question was issued on 13 March 13 2001, wherein Microcell announced its deployment of the enhanced 9-1-1 service in British Columbia and Alberta in conjunction with an application it had filed with its Part VII application to the CRTC requesting mandated provision of network access services for Wireless E9-1-1 in other parts of Canada (Microcell Telecommunications Inc., 2001g). Microcell's decision to issue the press release had apparently rocked the boat, and what had previously been an open discussion among the Wireless carriers and PSAPs turned into sharp disagreement over the appropriate timeline for deployment of Wireless E9-1-1 to customers. Ironically, the two incumbent and largest wireless carriers in the region, TELUS Mobility and Rogers AT&T Wireless, had drawn heavily on the symbolic image of a mobile phone as a personal safety device in order to market their products and services in Western Canada, yet they did not see fit to immediately provide Wireless E9-1-1 to their customers when was first made available.

Furthermore, this was not the first time that 9-1-1 had been introduced as a competitive issue among the wireless carriers. In fact, Clearnet had made mention of the CRTC's local competition framework to support its bid for a third-party 'Alternate Operator Services Provider' to handle Wireless E9-1-1 calls for the Ontario trial. Clearly the PSAP representatives took a dim view of the comments made by TELUS Mobility, as the minutes of the Working Group meeting for May 2001 note that the Ontario E9-1-1 Advisory board had 'expressed disappointment that once again 9-1-1 was being discussed as a competitive service' (Canadian Wireless Telecommunications Association, Wireless E9-1-1 Working Group, 2001d). The Ontario Board's position was not a new development either, as the board had previously submitted an appeal to the CRTC regarding the 'non-essential' status the Commission had conferred upon 9-1-1 emergency services in the Local Competition Framework decision in 1997 (Ontario 9-1-1 Advisory Board, 1997).

The Ontario trial was launched in June 2001 with an expected completion date of December. In many respects it was similar to the Alberta trial, with regional differences based on the Bell regional 9-1-1 system and the Toronto serving area and PSAPs. Bell began the trial fully intent on transitioning to a commercial rollout of Wireless E9-1-1 in Ontario. It would remain to be seen if there would prove to be a faster commercial uptake in Central Canada than there had been in Western Canada. Most certainly, Microcell would move to adopt Wireless E9-1-1 in order to meet its CLEC obligations, but would the other wireless carriers remain reluctant? Bell Mobility too, had by now moved into Western Canada and had suggested to the Working Group its intention to implement Wireless E9-1-1 in BC and Alberta sometime in 2002.

The record clearly shows that the wireless industry in Canada took the initiative to develop the technical capability for the first phase of Wireless E9-1-1 without a regulatory mandate to prod it on. Furthermore, two of the country's largest incumbent wireline carriers, TELUS and Bell Ontario, achieved commercial rollout to make the tariff available in most of the major population centres of Canada by 2002. What was disappointing, especially from the perspective of the PSAP representatives, was that the largely unregulated wireless carriers, with the exception of Microcell, appeared to be very reluctant and in some cases unwilling to deploy the service once it was made available.

Strategic Niche Management and the CWTA

I suggested earlier in this chapter that one possible strategy for mitigation-oriented policy is to encourage the introduction of third party, value-added service providers. The core idea is that these new entrants could play an important role in furthering experimentation on a broad scale to support innovation and expanded organizational competencies in the management of critical infrastructure. A review of the political economic perspective in this area, however, suggests that certain network elements may require regulatory supervision to ensure fair terms and conditions of access, as dominant players will tend to resist access to those elements as a means of protecting strategic interests in network interconnection arrangements. This suggests that where one finds contested or rejected problem formulations or design propositions among parties involved in a technology project, that it may be advisable to look closely for the possibility of strategic positioning behaviour among incumbents or other dominant stakeholders. From an analytical perspective, the point of such an inquiry is to identify typical network elements that may require more direct regulatory supervision in order to ensure that they remain accessible on fair terms and conditions to those interested parties who may wish to undertake innovative and experimental projects.

Innovation and Experimentation

The important role played by the CWTA Wireless E9-1-1 Working Group suggests that innovation in this kind of industry-directed forum may favour conservative problem-solving at the expense of permitting new entrants or high concept ideas that challenge current practices. The members of the Working Group, while forthright in their willingness to conduct technical trials for Wireless E9-1-1, were apparently uninterested in Clearnet's ambitious design proposition to introduce a third-party competitive 9-1-1 platform operator. Evidence from this case suggests that industry-directed innovation provides the basic foundation for strategic niche management but if it is not guided by carefully designed regulatory incentives, the process may allow a bias toward dominant or incumbent parties at the expense of innovative ideas and alternative approaches from new entrants. Such bias may be exercised through interconnection arrangements between service providers, and this is especially true with respect to key network elements that remain under the control of incumbent operators. In such cases, the cooperation (voluntary or otherwise) of these operators is essential if truly innovative design propositions are to be tested and refined.

Mark Armstrong reinforces the vital importance of network interconnection arrangements in achieving social objectives: 'A crucial issue for public policy is how to set the terms of access by rivals to ... monopolised inputs ('essential facilities') ...' (Armstrong, 1998, p. 545). This is often referred to as the bottleneck problem inherited from the pro-incumbent phase of interconnection policy. In the case of Wireless E9-1-1, the incumbent-owned and operated 9-1-1 platform is not deemed an 'essential facility' by the CRTC but nevertheless has proven to be a *de facto* bottleneck, as is especially notable in the case of Microcell's application for national wireless CLEC status (discussed in chapter four), where it was perhaps the centrally contested element of the proceeding. Even in the case of the cooperative technical trials, the incumbents played a dominant role in shaping problem formulation and dictating design propositions inasmuch as they effectively control the critical elements of E9-1-1 service delivery in each of their operating territories. The evidence available suggests that this situation conforms to what Armstrong calls a 'one-way access' situation:

> A characteristic of one-way access situations is that, for important inputs at least, regulation in terms of access is likely to be required to ensure socially-desirable outcomes in competitive sectors since the monopolizing firm will choose to set too high a level for access charges if free to do so. (Armstrong, 1998, p. 545)

While I concur with Armstrong's first point that terms and conditions of access are required to address the bottleneck problem, his point on access charges only touches upon one barrier to interconnection. In the case of Wireless E9-1-1 in

Canada, for example, access charges were not a prominent issue in the deployment of the service. On the contrary, there appeared to be considerable cooperation among all interested parties, particularly in the highly valued regional markets of Canada such as Alberta, BC, and Ontario. In other regions, however, cooperation from the incumbent operators has been far less forthcoming, as evidenced by Microcell's Part VII application to the CRTC. In such cases of incumbent resistance the issue was not over access charges but about access to technical information, existing technical capabilities in the 9-1-1 platforms, and concern over the choice and timing of equipment upgrade paths. For example, central to Microcell's Part VII application was a request to the CRTC to ensure that equipment upgrades in provinces with less advanced E9-1-1 systems, such as Manitoba and Nova Scotia, would be made in a timely manner and would be capable of accommodating the most common E9-1-1 solutions.

In this case, some of the indicators of strategic positioning introduced earlier in this chapter only became apparent in the upper layers of interconnection space, in part because of the high level of standardization and apparent consensus evident in lower layers of interconnection space. Two instances in the case seem to support the observation that attempts to gain strategic control may be shifting to the upper layers of interconnection space and that peripheral elements are now key areas of focus for dominant players. The first example is found in Bell's refusal to provide province-wide access to its Master Street Address Guide (MSAG) database when this was requested by Microcell as a required input to fulfill its controversial ALI-related obligations in the CRTC's Order 2000-831 (Bell Canada, 2001b). Microcell was considering how it could connect retail service points to the MSAG core network element in order to ensure that subscriber addresses could be pre-validated before being submitted for entry into the ALI database (Microcell Telecommunications Inc., 2001a). Access to the MSAG could be interpreted as an example of what Mansell termed a 'conditional access system.' The MSAG is provided by Bell to other carriers and is used to filter and verify street addresses listed on subscriber records when populating the ALI database. By setting the terms and conditions of access to the MSAG, a dominant player like Bell can exercise significant control over the development of Wireless E9-1-1. For example, Bell may have denied access to the MSAG simply to encourage Microcell to adopt interconnection arrangements more favourable to Bell.

The second example is found among the claims submitted by the incumbent carriers to the CRTC during proceedings on Microcell's Part VII Application. Incumbent carriers claimed that the release of information about PSAP equipment would harm the competitive position of Bell and other ILECs, and as such, it fell under section 39 of the *Telecommunications Act* (Aliant Telecom Inc. *et al.*, 2001b). The PSAPs objected to Bell's refusal to permit province-wide access to the MSAG, while Microcell criticized claims by the incumbents that information about PSAP equipment should be treated under section 39 of the *Act* (Microcell Telecommunications Inc., 2001d).

Viewing these concerns in terms of their relationship to interconnection space, the MSAG can be viewed as part of a value-added service to support the ALI function; and the inventories of PSAP equipment are located in the information services layer as a form of proprietary customer data (or so it was claimed by the incumbents). Both of these elements reside in the upper layers of interconnection space and at the edge of the CRTC's authority over interconnection arrangements. The incumbents, whether intentionally or not, used this regulatory uncertainty in the upper layers of interconnection space to influence lower level interconnection arrangements and to effectively delay the deployment of Wireless E9-1-1 in Canada.

Regulatory Intervention and Niche Management

The one-way access situation characteristic of the E9-1-1 bottleneck problem, which is evident other areas of network design, may be changing with the advent of new technologies and regulatory reform. Noam's forecasted modularity scenario implies that key network facilities may be distributed among a number of carriers, as well as unaffiliated third parties, requiring more complex *exchange* of services in order to create service packages for customers. The regulatory responsibility under such conditions is to determine, within a shifting environment of technologies and service design, what elements are essential and therefore deserve appropriate supervision. Interpretations of the definition of 'essential' also present a matter that might be influenced by a mitigation-oriented policy framework that seeks to promote greater experimentation and innovation in the management of critical infrastructure. In the case study, it is apparent that while 9-1-1 service does not fit the strict list of requirements to be an 'essential' service, in the Canadian local competition policy framework, it has been established as a *de facto* monopoly-controlled service through assertions made in the CWTA working group.

Regulatory intervention under such conditions is appropriate when there is reason to believe that dominant stakeholders are unduly influencing opportunities for innovation and experimentation, as seems to have been the case with the Clearnet proposal to introduce an alternative 9-1-1 platform operator, and similar proposals are also evident in the United States (TSI Telecommunication Services Inc., 2002). Had this design proposition been carried forward to a trial stage, the wireless carriers on the one hand may have found themselves in a more difficult and expensive position, needing to negotiate terms and conditions with more than one service provider to assemble a modular E9-1-1 system. On the other hand, however, such an arrangement could create greater room for experimentation and more rapid deployment of new services for public safety and commercial location-based services if it bypassed the efforts of incumbents to dictate the terms and conditions for interconnection with their 9-1-1 platforms. The modular scenario of service design may in fact become a plausible reality as wireless E9-1-1 enters into Phase Two development, where third party mobile positioning companies independent of the incumbent carriers enter to offer specialized location tracking services in conjunction with public safety.

The difficult job for policymakers and regulators is to identify the network elements that are or are likely to be most important for innovation and experimentation in the management of critical infrastructure. This job is made more difficult in an era of rapid technological change and substantial regulatory reform. Nevertheless, the core idea of intervention based on strategic niche management is that innovation and experimentation at the supply-side should be what CTA practitioners refer to as 'a probe and learn strategy' (Kemp, 1997). Policy must ensure that a suitable environment for probing and learning is maintained through fair terms and conditions of access to key network elements. Identifying those elements, however, is a very difficult challenge when faced with a design discourse characterized by multiple problem formulations and a diverse range of stakeholders. It is this challenge that is taken up in the next chapter.

Chapter 6

Communities of Experts

A year and a half after the successful completion of the technical trial in the Canadian province of Alberta and four months into a similar trial in Ontario, the CRTC was compelled to launch a public hearing on Wireless E9-1-1. It appeared that much of the easy success of early industry voluntary efforts had ground to a halt and that the longstanding policy of regulatory forbearance in the wireless sector might be called into question. The origins of CRTC Public Notice 2001-110, issued in October 2001, can be traced to the ambiguity of a single paragraph from the 1997 Local Competition Framework (Decision CRTC 97-8), out of which arose a seemingly intractable dispute between representatives from the telecom sector and those from the public safety agencies.

To comprehend the origins of this situation, it is necessary to again visit the wider regulatory context within which Wireless E9-1-1 in Canada has evolved and the series of events that led up to Public Notice 2001-110. This context also provides an opportunity to consider the role of the CRTC's Interconnection Steering Committee (CISC) from a technology assessment perspective as a 'locus for reflexivity,' different from both the strategic niche management function served by the CWTA's Wireless E9-1-1 Working Group and the technology forcing strategy employed by the FCC in the United States. While the CWTA was instrumental in fostering technical capabilities for Wireless E9-1-1 in Canada, the CISC by contrast played an equally important role in transforming it into a commercially available tariff through the design of network access services and by dealing with other lower layer interconnection issues. The CISC was also the site of a surprisingly intense row, however, between the wireless industry and the Public Safety Answering Point (PSAP) representatives over the issue of access to customer billing records. Within this debate there is clear evidence of multiple problem formulations as well as a revealing territorial conflict among the distinct communities of experts involved in this technology project.

Establishing Legitimacy

Inviting communities of experts to participate in a technology project requires some means of establishing their legitimacy as qualified participants. Legitimacy is very much part of the political and normative undertakings of Constructive Technology Assessment because it is essential to the matters of stakeholder representation and congruency. The events that led to the CRTC public hearing in 2001 provide an interesting study of the challenges to legitimacy that arise when communities of experts attempt to work together in an area without well defined boundaries or precedents. Challenges are likely to arise when policymakers adopt a strategy to expand stakeholder representation through loci for reflexivity.

The creation of loci for reflexivity is an intervention strategy 'that attempts to create and exploit *loci*: actual spaces, forums, and institutionalized linkages between supply and demand ... offering opportunities to modulate developments' (Schot and Rip, 1996, p.262). While CTA practitioners consider loci for reflexivity a strategy that coordinates interested parties from across both supply and demand sides of a technology project, under some circumstances this can be a murky distinction, given the variety of alignments and configurations of actors that appear at different layers of interconnection space. The distinction between this intervention strategy and others is perhaps more clearly understood if one recognizes that strategic niche management turns on building *congruency* (shared frames of reference) among stakeholders, whereas the reflexivity strategy pivots on the creation and maintenance of a *space of legitimacy* intended to extend the exchange of views on technology design than might otherwise be the case. This is not to say that congruency is not a key component of a reflexivity strategy, as CTA practitioners consider it to have an important role in fostering an expanded discourse and 'societal learning' through focussed debate (Grin and Graaf, 1996).

CTA practitioners have identified three types of *loci* where a reflexivity strategy might be attempted. The first of these is the 'action forum' such as the consensus conference and dialogue workshop. In most cases these are temporary loci and may be quite distant from the centres where technology development is actually taking place. Under such circumstances, feedback from the forums can be very limited and its actual influence in modulating technical design equally tempered. One North American example of this strategy can be found in the consensus conferences promoted by the Loka Institute in the United States (Sclove, 1999). The second type of *loci* is found where 'new platforms are created' to address emerging technologies. One example in the wireless industry is the Cellular Telecommunications and Internet Association (CTIA), founded as 'the voice of the wireless industry' with a mandate to undertake advocacy and information sharing on behalf of its membership. Typically, however, this kind of loci is more about removing barriers to new technology markets than expanding societal learning on technology issues and concerns. The third kind of *loci* is the institutional forum that has developed over time into well-established intermediaries for technology certification. One example is the Canadian Standards Association, which provides test labs and certification procedures for new products. Such organizations come with a caveat from the CTA practitioners, who suggest that within any given forum, or 'nexus' as it is also termed, learning may drift away from substantive debates about the wide social impact of a design proposition and toward issues of a comparatively benign bureaucratic nature:

> When a nexus has become institutionalized, learning shifts from the issue of how variation and selection can be linked, to learning how to handle specific technologies within the nexus. The institutionalization of the nexus makes it forceful, but it may

also create barriers to further broaden the design and development processes. (Schot and Rip, 1996, p. 262)

Like the intervention strategies themselves, such loci in actuality are usually a combination of types, performing various roles ranging from participatory action, to market cultivation, to an intermediary function. Against this backdrop of possibilities, the CRTC's Interconnection Steering Committee (CISC) can be regarded as an institutionalized locus for reflexivity based on a public-participatory and consensus-driven mandate. While the CRTC embodies tendencies toward the latter two concerns raised by CTA practitioners, particularly insofar as it is industry-dominated and bureaucratic, it also provides a forum for a wide scope of legitimate participants and presents a unique opportunity for undertaking technology assessment within a mitigation-oriented policy framework.

Seeking input from an expanded range of stakeholders can encourage learning, support demand articulation, generate alternative problem formulations, and lead to competing design propositions for assessment. Despite the possibilities, challenges abound, particularly with respect to establishing the legitimacy of participants in cases where communities of experts approach a problem from widely divergent perspectives. In order to understand this challenge more clearly this chapter begins by returning briefly to the literature on large technical systems to further consider the elements of growth and change in critical infrastructure.

Boundary Crossing

Jane Summerton has introduced the notion of 'boundary crossing' as a metaphor to interpret some of the challenges that arise when large technical systems go through periods of transformation (Summerton, 1994a, p. 6). Boundary crossing embodies a range of considerations that involve the social and technological factors that come into play when new stakeholders enter into contact with one another. Summerton has observed that such factors can be considered along functional, territorial, and cultural dimensions.

First, functional boundary crossing describes situations in which fundamentally different systems are interconnected, such as that between railways and communications infrastructures (Summerton, 1994a, p. 8). Such was the case with the convergence of the telegraph with the railways in the 19th century, where telegraph companies sought rights-of-way for their lines while their communications infrastructure provided a means to enhance railroad signaling capabilities. Second, territorial boundary crossing describes situations in which interconnection has jurisdictional implications of a political or organizational nature. Summerton cites the merging of the East and West German telecommunications networks as a case in point and she has noted that such events often begin with the disintegration of standing institutions and practices as a pre-requisite for an integration of separate systems. Indeed, this has been the case with regulatory reform in telecommunications and other infrastructures, where one of the first initiatives was to issue directives requiring the unbundling of monopoly-era networks as a step toward providing the functional integration needed to support the provision of services across competing carriers. Taken to its most extreme formulation, this unbundling process leads to Noam's modularity model, described in chapter three, where network elements are reduced to highly specialized functional units and assembled by third party systems integrators into client-specific applications. Third, cultural boundary crossing represents situations in which interconnection has an implication for normative values or practices among the interested parties (p. 10). These normative values may provide the hidden backdrop for subsequent decision-making and debate about technical details or jurisdictional control, suggesting to me that debates on matters of functionality and territoriality are more accurately described as two related subsets of cultural boundary crossing.

In other words, transformation in large technical systems involves a process of cultural or 'normative' boundary crossing that introduces issues related to function and territory for those parties involved. Any technology project, for that matter, will have elements of normative boundary crossing, particularly when distinct cultures of expertise are enlisted in an effort to achieve some sort of alignment toward a single design proposition.

Normative Perspectives on System Design

My previous point is supported by Janet Abbate's (1994) historical research on the debates in the early 1980s about the merits of data networking protocols TCP/IP and X.25. Abbate's findings provide an insightful perspective on the evolution of internetworking standards and support the view that challenges to boundary crossing may be, in essence, a matter of normative debate, in some instances with territorial concerns expressed through functional claims. She contends, for instance, that conflict over competing design propositions may be only partly about technical matters and a good deal more about group interests and 'fundamental differences in [stakeholder] perceptions of the technology' within a wider social setting (p. 194).

Abbate examined the historical development of the data networking protocols TCP/IP (datagram-based) and X.25 (virtual circuit-based) in early internetworking initiatives between 1976 and the late 1980s. The term 'TCP/IP' refers to a datagram-based protocol capable of flexible routing over multiple network pathways and it is usually associated with public use of the Internet. The term 'X.25' refers to a lesser-known networking protocol that provides the backbone for numerous private data networks around the world. According to Abbate's account of the development of these protocols, the internetworking debate taken up by the engineers largely involved technical claims about the functional merits of each standard. On closer examination, however, the claims revealed themselves as manifestations of normative perspectives on network design.

At the heart of the debate and the veiled normative perspectives was a basic question about the appropriate site of control in a data network. Datagram-based networks using TCP/IP place greater control at the edge and upper layers of the network, in the host computers, whereas virtual circuit-based designs using X.25

place control at the lower layers of interconnection space, among the nodes and transmission media at the core of the network. Looking beyond the complex and often contradictory debates about the various technical merits of these two standards, Abbate points out that each standardization initiative represented a very distinct school of thought in the field of data networking: TCP/IP was a product of the U.S. Department of Defense's ARPANET, while X.25 emerged from international consensus with the CCITT (Consultative Committee on International Telegraphy and Telephony) of the International Telecommunications Union (ITU). Abbate suggests that such institutional connections are important factors, particularly where '[a]rguments tend to be framed as the evaluation of trade-offs between different technical costs and benefits, yet the assessment of these trade-offs *depends on tacit assumptions about the goals and operating environment of the networks*' [emphasis added] (Abbate, 1994, p. 199). Her choice of the term 'tacit assumptions' suggests that within any given technology project, problem formulations expressed by interested parties will have extensive normative dimensions that may be significant yet unacknowledged in debates that centre on the technical merit of competing design propositions.

In the case of internetworking standards, these assumptions were articulated, on the one hand, through debates about the merits of TCP/IP and X.25 as technical standards. On the other hand, upon closer scrutiny of the evaluative criteria used by each of the camps, Abbate found that each tended to reflect different baseline assumptions about what constitutes appropriate network design. More specifically, these assumptions appeared to correspond to the organizational setting from which the parties were affiliated. For example, European participants in the X.25 camp viewed network design from within an institutional culture that envisioned public networks as tightly controlled, vertically integrated structures dominated by state monopolies. As a result, these participants held the view that a central body should administer overall network design and transactions with control over nodes and transmission, using gateways circumscribed by political boundaries, and X.25 was well suited to this design. The American participants, however, viewed network design from a rapidly changing environment dominated by computing specialists and characterized by the dismantling of AT&T, the rise of competing network providers, and a well-integrated North American telecommunications infrastructure. From their perspective, growing heterogeneity in data networks appeared to be the normal course of development and thus control was best left to the edges with the host computers. For the Americans, TCP/IP was better suited to this context, rather than an attempt to muster a centrally managed network architecture.

Abbate summarizes the difference in perspectives: 'At the heart of the arguments over [TCP/IP and X.25] are two contrasting assumptions about the environment in which large networks will operate' (1994, p. 204). Each of the data networking standards emerged out of and tended to support regionally defined and organizationally congruent perspectives. Abbate concludes with the recommendation that observers must therefore consider such normative reference points when seeking 'to identify where the gulfs lie between competing technologies and to understand the strategies of those who attempt to cross them' (1994, p. 208).

Technological Frame

An important question for mitigation-oriented policy research emerges from Abbate's findings, specifically with respect to establishing a link between dynamic pressures and root causes of risk and vulnerability. Whereas functional and territorial dimensions of boundary crossing stem from dynamic pressures such as rapid technological change or regulatory reform, the normative dimension is more indicative of root causes linked to economic, cultural, or political settings. This begs the question of the relationship between root causes and dynamic pressures. Or put another way, how is it that functional and territorial claims come to be associated with normative perspectives, and in turn, how might these come to be expressed by interested parties in debates over design propositions? The question is particularly relevant when policymakers are formulating a strategy based on loci for reflexivity, where different cultures of expertise are deliberately enlisted in a technology project. Answers to the question may provide insight as to the possible kinds of obstacles one may encounter in an expanded design discourse.

Bijker's concept of the 'technological frame' is a useful means to get at this relationship. He first introduced it to describe the basis for interaction between and among interested parties, suggesting it be 'used by the analyst to order data and to facilitate the interpretation of the interactions within a relevant social group' (Bijker, 1995, p. 122, 124). At a rudimentary level, the technological frame is something like a shared perspective manifested as commonly held goals, problems and problem solving strategies, theories, tacit knowledge, and practices. It provides a context that actively influences demand articulation, problem formulation and design propositions, thereby enabling and constraining criteria of acceptability applied in a technology assessment. Bijker, however, is careful to define a technological frame as the outcome of situated interaction and not an *a priori* property of individuals or social groups *per se*:[1]

> A technological frame structures the interactions among the actors of a relevant social group. Thus it is not an individual's characteristic, nor a characteristic of systems or institutions; technological frames are located between actors, not in actors or above actors. A technological frame is built up when interaction 'around' an artefact begins. Existing practice does guide future practice, though without logical determination. If existing interactions move members of an emerging relevant social group in the same

---

[1] Admittedly this remains to be further examined from a theoretical standpoint. In my view, Bijker's formulation is similar to that of actor network theory, especially with its similarities to semiotics and Symbolic Interactionism. See, for instance, Law's perspective on this matter (Law, 1999). For this study, however, it nevertheless provides a good foundation (as Bijker himself notes) for ordering raw data and enhancing an interpretation of contextual factors that influence a technology project.

direction, a technological frame will build up; if not, there will be no frame, no relevant social group, no future interaction. (Bijker, 1995, p. 123)

Others have found Bijker's concept useful to interpret the effects of technological change within organizational settings. Orlikowski and Gash, for instance, take the idea of a technological frame to be a subset within a wider field dealing with social cognition. More specifically, they consider Bijker's concept a 'contextual dimension' of social cognition, defining it as:

> ... the understanding that members of a social group come to have of particular technological artifacts, and they include not only knowledge about the particular technology but also local understanding of specific uses in a given setting. ...
>
> Technological frames have powerful effects in that people's assumptions, expectations, and knowledge about the purpose, context, importance, and role of technology will strongly influence the choices made regarding the design and use of those technologies. (Orlikowski and Gash, 1994, p. 178)

In sum, the technological frame describes a normative boundary, demarcating various social groups and lending character to their perspectives on a technology project. Bijker claims that 'a technological frame comprises all the elements that influence the interactions within relevant social groups and lead to the attribution of meanings to technical artifacts' (1995, p. 123). The following table is Bijker's 'tentative list' of elements in a technological frame, with the addition of two corresponding columns—'Activity' and 'Analyst's question(s)'—intended to complement empirical research efforts.[2] (See Table 6.1)

Bijker's Configuration Model

Having introduced his readers to the technological frame idea, Bijker then offers a model to put it into practice, claiming that in a given technology project at least one of three situations are common: an absence of a technological frame, one dominant frame, or competing technological frames (Bijker, 1995, p. 276-279). He describes this set of arrangements as a 'configuration model' of technological change, for which he then posits a set of hypothetical expectations. Bijker writes, for instance, that 'where there is no dominant technological frame, the range of variants that can be put forward to solve a problem is relatively unconstrained' (p. 277). This suggests that where there is the absence of a technological frame an observer will typically find a wide variety of problem formulations and design propositions put forward by interested parties.

---

[2] Bijker does not provide these extra elements. I have derived them from findings he presents in a case study of celluloid chemists involved in the creation of Bakelite. See Bijker (1995, p. 126).

Table 6.1 Elements of Bijker's Technological Frame

Element	Activity	Analyst's question(s)
Goals	Demand articulation	What motivates the technology project?
Key problems	Demand articulation	What issues are being addressed by the technology project?
Problem-solving strategies	Problem formulation	What are the identified general principles to solving the problem?
Requirements to be met by solutions	Problem formulation	What are the standards by which design propositions are to be measured?
Current theories	Problem formulation	How is the problem conceptualized?
Tacit knowledge	Problem formulation	What are the taken-for-granted assumptions?
Testing procedures	Design proposition	What is an acceptable means for evaluating a design proposition?
Design methods and criteria	Design proposition	What existing practices are drawn upon to establish a design proposition?
Users' practice	Problem formulation/Design proposition	What are the views on users' needs and behaviours with respect to a design proposition?
Perceived substitution function	Design proposition	What are the relevant alternatives to a design proposition?
Exemplary artifacts	Problem formulation/Design proposition	What previously established artifacts or systems reside at the heart of a design proposition?

Where we find a single dominant frame and a set of vested interests, Bijker indicates that interested parties will take a much more conservative approach to problem formulation, typically dominated by a single group, thereby leading to greater conformity in problem formulation and design propositions. In these instances, Bijker introduces the term 'inclusion' to describe the variations between relevant social groups that are more or less within the dominant frame and those that are outside of it. Those within the frame ('high inclusion actors') are likely to draw on standard problem-solving strategies based on an exemplary artifact or system because,

> [t]he relevant social groups have, in building up the technological frame, invested so much in the artifact that its meaning has become quite fixed—it cannot be changed

easily, and it forms part of a hardened network of practices, theories, and social institutions. From this time on it may indeed happen that, naively spoken, an artifact 'determines' social development. (Bijker, 1995, p. 282)

Those actors exhibiting 'low inclusion' may offer alternative approaches by disregarding or otherwise questioning the exemplary artifact/system within the dominant technological frame, or perhaps complying with them as 'obligatory passage points' towards a desired objective (p. 285).

In the case of multiple and equally dominant frames, the configuration model proposes that 'arguments, criteria, and considerations that are valid in one technological frame will not carry much weight in the other [competing] frames' (Bijker, 1995, p. 279). Under these circumstances, criteria external to all technological frames will come to play a major role in problem formulation and evaluation of design propositions. Bijker suggests that in such cases alternative rhetorical strategies and an 'amalgamation of vested interests' (p. 279) may come to play a significant role in proceedings.

Within Bijker's configuration model, the technological frame serves as a normative boundary that embodies territorial and functional dimensions. First, it establishes a territory by which various social groups are granted legitimacy (high or low inclusion) within a technology project and in that manner, a technological frame can provide a position from which to attack or defend various problem formulations or design propositions. Bijker's configuration model also sets out expectations for those situations in which territorial boundaries come into conflict—that is, when dominant or competing technological frames are at play. Second, it provides the context to evaluate functional claims concerning the relative merit of a given design proposition. For example, in Abbate's case of internetworking standards, the data networking protocols TCP/IP and X.25 served as exemplary artifacts within two competing technological frames, each with a different basis for assessing network control architecture. These two frames, however, were largely comprised of more extensive normative positions arising from regional contexts and institutional perspectives between Europe and America, and between the distinct cultures that exist within the worlds of telecom engineering and computer networking.

Boundaries and Critical Infrastructure

This idea of boundary crossing emphasizes the point that growth and change in large technical systems introduces new alignments among interested parties and that this process of change is laden with normative values. In Summerton's words, 'the rhetoric of defining the new system (what is possible, what will work) can be as important as actual system design' (Summerton, 1994a, p. 16). An intervention strategy based on loci for reflexivity is a coordinated effort to expand the range of 'rhetoric' in which network design exists and thus brings with it the prospect of new forms of boundary crossing and concomitant challenges for achieving viable outcomes. Bijker's technological frame and configuration model provide a way to study boundary crossing incidents as a means of improving intervention strategies for the management of critical infrastructure.

Whereas the first and second facets of the Wireless E9-1-1 case examined standardization and innovation efforts, the third facet of the case considers the influence of normative perspectives on the communities of experts that were involved in this public safety telecommunications initiative. In particular, this aspect of the case draws on the boundary crossing theme and presents an interpretation of the dispute that surfaced in the proceedings around a controversial design proposition involving the ALI database. An analysis and understanding of such disputes is important for mitigation-oriented policy research because it may provide some measure of guidance in planning and executing an effective intervention strategy that seeks to expand stakeholder participation in the management of critical infrastructure. More specifically, an evaluative framework that takes into consideration the active role of technological frames makes possible an expanded technology assessment that includes contextual explanations for conflict situations that perhaps are not adequately accounted for by political, economic or purely technical perspectives. In addition, the configuration model offers some insight as to the rhetorical strategies that parties adopt when seeking to promote design propositions. Such findings are important for policymakers who may attempt to establish loci for reflexivity as a strategy to improve the interaction of stakeholders with a diversity of backgrounds from multiple domains of expertise.

The Disputed Decision

In May 1999 Microcell Connexions filed a general tariff notice with the CRTC and declared its intention to become a Competitive Local Exchange Carrier (CLEC) in a number of Canadian regions (Microcell Connexions Inc., 1999). Microcell, a Montreal-based wireless carrier holding a national digital PCS license in Canada, offering mobile telephone services under the brand name Fido, had announced a year prior to its formal application to the CRTC that co-carrier status was 'an integral part of Microcell's business plan and vision for the future of wireless' (Microcell Connexions Inc., 1998). In December 1999, however, the CRTC issued Telecom Order 99-1127 denying the Microcell tariff application on the grounds that is was 'inconsistent with the interconnection framework that underpins local competition in Canada' and that the tariff 'is significantly different from the tariffs of the incumbent local exchange carriers (ILECs) and CLECs' (Canadian Radio-television and Telecommunications Commission, 1999).

It is interesting to note that Order 99-1127 makes no mention of E9-1-1, as the CRTC appeared to be concerned solely with Microcell's apparent failure to adequately address other regulatory matters such as equal access to long distance providers and customer privacy. Microcell had tried in advance to address a number of these shortcomings by arguing that they stemmed directly from two factors that set Microcell operations apart from other existing CLECs and ILECs—

namely, that Microcell would operate as a wholesaler of telecom services and that the Microcell *mobile* wireless network differed both in functionality and structure from conventional wireline networks. Despite the fact that Canada's local competition policy had been framed around the objective of technology neutrality, the matter of Microcell's proposed wholesale operation and its mobile network infrastructure were both to become central factors in a technological framing exercise to interpret the E9-1-1 obligations, which had been established in paragraph 286 of the CRTC's local competition framework in Decision 97-8.

Microcell filed a revised general tariff application the following year in May 2000, with its national competitor Clearnet PCS filing for similar CLEC status a month later (Clearnet PCS Inc., 2000; Microcell Connexions Inc., 2000a). This time around Microcell's application received interim approval along with the Clearnet application in CRTC Orders 2000-830 and 2000-831, issued in September (Canadian Radio-television and Telecommunications Commission, 2000a, 2000b).[3] Clearnet's application was placed in abeyance when the incumbent operator TELUS acquired it in October 2000 and national operations were subsequently undertaken by TELUS Mobility. While the Clearnet application in effect had been suspended, Microcell continued to forge ahead with its plan to become a CLEC. CRTC Order 2000-831 set out a number of obligations for Microcell, including a controversial directive concerning E9-1-1 that eventually forced the regulator to become directly involved in the public safety facet of the matter when it issued its public call for input in Public Notice 2001-110.

To grasp the full context within which the CRTC Orders 2000-830/831 were issued, it is essential to note that the successful Alberta wireless technical trial had ended a month prior to Microcell's original application in 1999 to become a CLEC. The Alberta trial had proven the feasibility of Phase One Wireless E9-1-1 in a major Canadian operating territory and it had established an accepted design based on the provision of Emergency Service Routing Digits (ESRD) and 10-digit ANI to the Public Safety Answering Points (PSAPs). Drawing on the success of the Alberta trial, at the time of Order 2000-831, incumbent TELUS was also preparing a tariff application for the deployment of Wireless E9-1-1 network access services in the provinces of Alberta and British Columbia. It was therefore to come as a surprise to many involved when the CRTC in Order 2000-831, issued the following directive to Microcell regarding its 9-1-1 obligations as a Wireless CLEC:

> The Commission considers that, as a CLEC, Microcell should provide the end-users of its resellers with 9-1-1 service that is better than what it currently provides them as a WSP [Wireless Service Provider]. The Commission considers that until Wireless E9-1-1 is implemented, Microcell *should support the inclusion of the subscriber records of its resellers' end-users in the ALI [automatic location identification] databases.* [emphasis added]

---

[3] Order 2000-830 addressed the Clearnet PCS application, while Order 2000-831 addressed the Microcell application.

The Commission [hereby] directs Microcell to update the relevant ALI databases with the subscriber records of its resellers' end-users where it operates as a CLEC... (Canadian Radio-television and Telecommunications Commission, 2000b, par. 51, 52)

The CRTC's directive stunned both wireless carriers and incumbent wireline operators alike. The very fact that there had been no discussion or attempt to populate the ALI database with mobile subscriber records within the CWTA's technical trials only added to the confusion and dismay, most especially for Microcell. The wireless carrier had previously attempted to address the problematic wording of paragraph 286 in Decision 97-8 with its CLEC application to the regulator, which noted that a literal interpretation of the E9-1-1 obligation as worded in paragraph 286 was problematic for would-be wireless CLECs:

With regard to the provision of appropriate end-user information to the Automatic Location Identification (ALI) database, we note the means by which this can or should occur is different for a mobile wireless carrier than for a fixed wireline carrier. In the case of a fixed wireline carrier, a permanent data file is entered into the ALI database for each end user, and consists of the end user's name, fixed service address and 7-digit local telephone number. However, in the case of a mobile wireless carrier, such data entry protocol clearly would be inappropriate.

Mobile wireless subscribers do not remain at their fixed billing address, nor can they be uniquely identified by a 7-digit local telephone number. (A 10-digit phone number is required due to the fact that mobile wireless subscribers can roam outside their home NPA). As such it would be inappropriate and potentially confusing to enter mobile wireless end user information into the ALI database in the same manner as for fixed wireline subscribers. (Microcell Connexions Inc., 2000a)

The wording of paragraph 286 nevertheless remained a problem for Microcell and its CLEC application. Microcell even made reference to the successful Alberta trial and to the likelihood that TELUS would file a tariff proposal in the near future, noting further that incumbent Bell in Ontario and Quebec, and MT&T in Nova Scotia were actively working toward making Wireless E9-1-1 available through a similar series of technical trials in their respective operating territories. Microcell's application suggested that there was a congenial and progressive atmosphere among the various stakeholders, a sense conveyed in its expressed opinion that 'a migration to Wireless E9-1-1 can and should occur swiftly across Canada.' In the meantime, however, Microcell admitted that wireless carriers would 'have no practical choice but to continue to employ existing 9-1-1 call routing arrangements,' which meant line-side interconnection with no enhanced functions. Microcell's commitment was founded on what it called a 'trial-before-tariff' argument, which advocated that technical trials be undertaken in each of the ILEC serving territories as a prerequisite to filing commercial Wireless E9-1-1 network access service tariffs. In sum, Microcell had asked the CRTC to maintain the status quo in the interim but added a pledge to adopt Wireless E9-1-1 service as it gradually became available on a commercial basis across the country. Clearnet's tariff application presented a largely identical argument.

Comments submitted to the CRTC in response to Microcell's tariff application came from TELUS, Bell and its affiliates, and the Alberta E9-1-1 Advisory Association (AEAA) on behalf of the public safety answering points across Canada. Incumbent carrier Bell and the AEAA objected to Microcell's deployment strategy based on the view that arrangements could be made immediately available to improve the functionality of wireless 9-1-1 before the implementation of a commercial tariff. Bell stated that it was prepared to make a trunk-side interconnection arrangement available that would provide some enhanced features to Microcell. Bell was also concerned that Microcell's proposed 'hybrid' solution, as Bell referred to it, would establish an asymmetrical regulatory environment in which wireless CLECs could circumvent various obligations on the basis of technical arguments (Bell Canada, 2000a). In a similar set of comments in response to the Clearnet tariff notice, Bell stated that 'Agreeing to treat wireless CLECs differently to any other established CLECs might pose serious issues, concerns and general inequities about emerging [wireless] technology-based [local exchange carriers]' (Bell Canada, 2000b, par. 28).

The AEAA, speaking on behalf of the public safety answering points (PSAPs), characterized the advent of mobile wireless technology as an 'ongoing erosion of the E9-1-1 database' and objected to Microcell's interpretation of paragraph 286 by claiming that 'the inclusion of any and all information that is automatically delivered with an E9-1-1 call is vital' in order to fulfil 'the PSAPs legal and moral obligation to locate callers who call 9-1-1 but can't communicate ...' (Alberta E9-1-1 Advisory Association, 2000b). TELUS was largely silent on the matter of Wireless E9-1-1 design, choosing instead to criticize Microcell's reluctance to enter into service contracts prior to the approval of commercial tariffs for the new E9-1-1 service (TELUS Corporation, 2000). Most importantly, a number of parties to the proceeding suggested that Wireless E9-1-1 interconnection arrangements be moved to the CRTC's Interconnection Steering Committee (CISC) Emergency Services Working Group rather than continuing with the industry working group under the CWTA, where the technical arrangements had first been developed and successfully tested.

Microcell's response to these suggestions was that its own proposed solution to the Wireless E9-1-1 deployment problem was the most effective means 'to minimize disruption and confusion' among the various parties during the transitional period. In other words, Microcell wanted to avoid the deployment of *ad hoc* and non-standard solutions in favour of a more stable, long-term approach to solving the problem of Wireless E9-1-1. Furthermore, Microcell suggested in its reply comments that Bell's offer of an interim interconnection arrangement was 'greatly oversimplifying the effort required to implement even a partial enhancement of wireless 9-1-1,' citing the findings of the Alberta trial report as an example of the extensive consultation and preparation required for proper deployment (Microcell Connexions Inc., 2000b, par. 47).

The public safety agencies, through the AEAA representative, recommended that as an interim solution in operating territories without Wireless E9-1-1, that Microcell be required to populate the ALI database with its customers' billing records. Microcell dismissed this idea on the grounds that these records were not useful for locating customers who might be calling 9-1-1 on their mobile phone and moreover, that this design proposition had not formed any part of the Wireless E9-1-1 discussion within the scope of industry activities or for that matter, within the FCC mandate in the United States on which the Canadian design was based. In addition, Microcell claimed that the PSAPs were not technically capable of automatically receiving the data even if it were made available, with the wry suggestion that another agenda might underlie the PSAP urgings on the matter:

> Microcell notes that the law enforcement community has raised the issue of the entry of wireless subscriber records into the ALI database as part of a broader effort to create a universal telecommunications subscriber information database in Canada. This effort has been met with expressions of serious concern from wireline and wireless carriers alike. Microcell suggests that if the PSAP or law enforcement communities want to pursue this issue further, it should not be in the context of a Microcell tariff notice but rather in the context of an explicit and open application to the Commission. (Microcell Connexions Inc., 2000b, par. 74)

Microcell repeated this allegation within the CISC Emergency Services Working Group in February of 2001, pointing out that nowhere in previous trials had customer billing records ever been considered as a feature of Wireless E9-1-1. The idea, according to Microcell, may have come from the CISC Network Security Working Group, 'where law enforcement representatives had included it as part of an omnibus proposal to develop what could best be referred to as a universal telecommunications subscriber information database covering both wireline and wireless subscriber data.' Microcell further suggested, 'the fact that this initial discussion took place in a law enforcement context raises questions as to the relative weight of 9-1-1 service motives and non 9-1-1 service motives behind the [ALI] proposal' (Microcell Telecommunications Inc., 2001f, sec. 2.3).

The need to minimize unnecessary confusion and disruption was again cited by Microcell to defend the CWTA as an appropriate forum for the continued development of Wireless E9-1-1, as opposed to introducing the CISC into the process as had been recommended by the ILECs and AEAA. One possible clue to this objection may be derived from comments made by Microcell in its CLEC application, where it made reference on a related matter to delays inherent in the CISC process and noted that, 'We believe there are incentives for the ILECs, and perhaps other parties, to extend that delay' (Microcell Connexions Inc., 2000a, schedule 1, section E).

As it turned out, the CRTC ruled largely in favour of the comments submitted by the ILECs and the AEAA. Microcell was to implement Wireless E9-1-1 where and when it became technically feasible, but in the meantime, Microcell would be required to populate the ALI database with customer billing records and would adopt Bell's temporary interconnection arrangements in order to provide a limited enhanced functionality in other operating territories. Furthermore, and also in keeping with recommendations of the ILECS and the AEAA, the CISC Emergency Services Working Group (ESWG) was also directed by the CRTC to become involved and to develop the necessary interconnection arrangements for Wireless E9-1-1.

By February 2001, TELUS had received interim approval from the CRTC for its Wireless E9-1-1 network access service. This tariff enabled Microcell to begin offering Wireless E9-1-1 in Alberta and parts of British Columbia in March 2001 (Microcell Telecommunications Inc., 2001g). With regard to the CRTC's directives in Order 2000-831, the technical details of populating the ALI database with customer billing records were now being discussed formally in the CISC, as was a revision of the 'CLEC Trunk-Side Interconnection Document' that would provide the vehicle of inscription necessary for standardizing the provision of E9-1-1 network access services to wireless CLECs.

Deployment in Western Canada

The final step toward the deployment of Wireless E9-1-1 in Canada was taken on 8 December 2000, when incumbent TELUS filed two tariff notices with the CRTC—identified as 'TN 327' and 'TN 4120'—both titled 'Wireless Service Provider Enhanced Provincial 9-1-1 Network Access Service' (TELUS Communications (B.C.), 2000; TELUS Communications Inc., 2000). These filings seemed to align parties to the Wireless E9-1-1 technical trials into a stable socio-technical system through a set of related vehicles of inscription. The first of these vehicles of inscription was the tariff notice itself, which is a statement of intent made available for public consultation, that sets out the terms and conditions of a proposed telecommunications service offering. The second and third vehicles of inscription are the actual revisions to the ILEC's tariff page and the set of legal agreements between the ILEC, the wireless carrier, and the local municipality or PSAP. Since the Wireless E9-1-1 system is designed in two legs, there is a requirement for documents that deal with the first leg (wireless CLEC to ILEC) and documents for the second leg (ILEC to PSAP and wireless CLEC to PSAP). As such, the TELUS Wireless E9-1-1 service offering consisted of a Tariff Notice, revisions to its current E9-1-1 service offering, a definition of Wireless E9-1-1 service in the General Tariff, as well as corresponding legal agreements.

The reply comments that followed the TELUS tariff notice included submissions from the Alberta E9-1-1 Advisory Association (AEAA), the BC 9-1-1 Service Providers Association (BCSPA), and the Canadian Wireless Telecommunications Association (CWTA). Once again, the PSAP representatives raised the issue of wireless customer billing records and the ALI database. TELUS responded with the following comments to the regulator:

> The proposed service is the result of the findings and conclusions reached during the Alberta Enhanced 9-1-1 Wireless Trial, the scope of which was understood and agreed to by the members of the industry who participated in the trial. The trial and subsequent service development did not take into account the technical and operational issues or the costs associated with the management of such an enhancement to the proposed service. The AEAA's request [to include customer billing records in the ALI database] is therefore clearly beyond the scope of the proceeding surrounding TN 327

and TN 4120, a fact also noted by the CWTA in its comments, and should be rejected. (TELUS Communications Inc., 2001b, par. 6)

Despite the fact that CRTC Order 2000-831 supported the position of the AEAA on customer billing records and the ALI database, comments issued by the British Columbia 9-1-1 Service Providers Association (BCSPA)—an organization that had originally aligned with the AEAA's position—were revised to more closely align with the CWTA position and in so doing, split the position of the PSAPs on this matter. Comments issued by the BCSPA were not to be taken lightly, as they represented the first municipal region in Canada that was in point of fact prepared to deploy Wireless E9-1-1 (BC 9-1-1 Service Providers Association, 2001b).[4]

In its comments to the CRTC, the AEAA also included a statement of 'strong support' for the proposed tariffs in a letter of thanks sent to TELUS in June 2000, after the completion of the Alberta trial.. Oddly enough, even though the AEAA appeared to have a firm grasp of the proposed design, the letter made no mention of wireless customer billing records in the ALI database. The AEAA comments also introduced a line of argument that would persist in other proceedings, yet the comments did not deal with the inherent problem that mobile terminals presented to an E9-1-1 service. In its comments, the AEAA suggested that even with high resolution location capabilities (i.e., FCC Phase Two design), Wireless E9-1-1 systems 'would still only get emergency services to the front door of an apartment building' and that '[p]roperly provisioned "wireless service provider ALI records" [*sic*] in the database, delivered concurrent with the voice call will guide them to the correct apartment in a precise and timely fashion.' (Alberta E9-1-1 Advisory Association, 2001, par. 7).[5]

Of course the problem with this line of argument is that it assumes mobile calls are likely to be made from the same address as those listed in customer billing records. Such is the case with traditional wireline E9-1-1, where the telephone is usually fixed to a single location. Mobile telephones—as Microcell would later go to great lengths to illustrate to the CRTC—do not conform in any reliable sense to the original ALI administrative system, hence the need for an alternative design.

---

[4] In their comments to the CRTC, the BCSPA first aligned directly with the AEAA in requesting that all WSPs be required to place their subscriber records in the ALI database. A revised version of the comments (issued later the same day) contained the following clarification vis-à-vis their original position: 'My intention is to stress that all wireless service providers be made *to provide the service being proposed by TELUS in this Tariff application.*' [emphasis added] (BC 9-1-1 Service Providers Association, 2001b). Effectively, this revision advocated the design of the TELUS system without forcing an outright dismissal of the ALI directive, but the message clearly seemed to be that the ALI requirement was not of significant interest to the BCSPA in this phase of Wireless E9-1-1 deployment. Personal conversations I have had with BCSPA representatives confirmed this to be the case.

[5] The Ontario E9-1-1 Advisory Board also used the 'apartment door' argument in a Contribution document to the CISC Emergency Services Working Group in March 2001 (Ontario 9-1-1 Advisory Board and Alberta 9-1-1 Advisory Association, 2001b).

While it is conceivable that wireless customer billing records could indeed be useful under certain circumstances, the AEAA wanted *real-time access* to subscriber records that were 'delivered concurrent with the voice call,' a design requirement that had been dismissed repeatedly by the CWTA and the ILECs as either technically or administratively infeasible.

One of the appendices to the AEAA comments is interesting in this respect because it sheds some insight on the organizational culture of the public safety agencies and the manner in which they interpret their obligation to serve the public. The letter was written by the Deputy Fire Chief of Strathcona County in Alberta (it is undated and was originally addressed to the CRTC), and is titled 'Consequences Caused by the Lack of Wireless Callers [*sic*] Information.' It describes a murder that took place in a rural part of central Alberta in late August 1999, where the lack of ANI/ALI capability was blamed for an otherwise unnecessary delay in the arrival of emergency services at the scene of the crime. According to the account, three people died on the scene because the emergency operator was unable to provide specific location information to emergency services personnel. The author of the letter admonished the incumbent operator TELUS for its public safety-based marketing strategy—'Cellular Saves Lives' was its slogan at the time—claiming the marketing messages were 'dangerous' because of the false expectations created in mobile phone customers. Perhaps the most significant contribution made by this letter to the Wireless E9-1-1 case, however, was that it introduced an otherwise overlooked human interest angle to the situation. The author expresses the personal frustration that he and his staff appear to have experienced repeatedly with emergency calls originating from mobile phones. He also described the emotional impact the incident had on the operator who handled the call:

> The Dispatcher answering the call for help from the lady being shot has suffered through a very traumatic experience. Not having basic information gave her a sense of not being able to help the caller during her last few minutes of life. Further to this, talking to her husband [*sic*] when he knew his wife was dead and venting his frustrations over the phone. This call has had a profound impact on the Dispatcher. The County will endeavour to seek ... counselling for the individual, however, I believe the Dispatcher will be scarred for life by this call. (Alberta E9-1-1 Advisory Association, 2001, Attachment C)

Testimony given in this letter reflects a fundamental difference in problem formulations between the wireless carriers, the ILECs and the PSAPs. Whereas the CWTA referred to a need to minimize costs and Microcell advocated a need to minimize confusion and disruption in the transition to Wireless E9-1-1, the PSAPs formulated the problem according to its impact on professional practice, their legal and moral obligations to serve the public, and its ultimate effect on human lives.

Other issues introduced during the Tariff Notice proceeding included concern about liability and indemnity provisions in the various agreements between wireless carriers, PSAPs, and incumbents. For instance, the CWTA submitted comments to the effect that the legal agreements were one-sided and provided unfavourable terms for TELUS on the matters of liability and indemnity. TELUS responded with the claim that the legal agreements were based on a previous wireline E9-1-1 agreement established through the CRTC Interconnection Steering Committee and therefore provided acceptable terms and conditions (TELUS Communications (B.C.), 2000).

Wireless E9-1-1 deployment under the TELUS tariff finally received interim approval by the CRTC on 2 February 2001 (Canadian Radio-television and Telecommunications Commission, 2001g). Shortly thereafter, Microcell introduced the service to its customers in the provinces of Alberta and British Columbia.

Deployment in Central Canada

At the other end of the country, similar steps were being taken toward the deployment of Wireless E9-1-1. After the conclusion of the Ontario Wireless Trial in November 2001, and a year after the TELUS filing, incumbent Bell filed its own Tariff Notice for Wireless E9-1-1 network access service (Bell Canada, 2001a). Bell's stated intent was to make the service available in the Toronto region by January 2002, with wider deployment planned for June. In the public proceeding that followed, comments from the wireless industry were not submitted by the CWTA but rather by Rogers Wireless Inc. and Microcell individually. Microcell largely applauded the tariff proposal and recommended only minor changes, but did recommend that the CRTC direct Bell to produce a firm deployment schedule for the service. Microcell did request certain clarifications, first, on rates and charges pertaining to Bell's status as a wireless CLEC and second, on whether Bell would be required to pay the full CLEC rate for E9-1-1 service or whether the half-rate for wireless carriers established in Decision 99-17 would remain in effect (Microcell Telecommunications Inc., 2001e, par. 5, 14).[6]

During this proceeding, the wireless carrier Rogers Wireless raised numerous concerns with respect to liability and indemnity provisions in the legal agreement, claiming that they were 'biased in favour of Bell' (Rogers Wireless Inc., 2001, par. 4). Rogers Wireless also expressed more specific concerns with items such as trunking arrangements, telephone number charges, termination conditions, and signalling facilities, most of which had been raised by the CWTA in the TELUS tariff proceeding. Rogers Wireless characterized these concerns as issues of regulatory asymmetry insofar as the Bell tariff would exert unfair terms and conditions on the wireless carriers. Above all, however, Rogers Wireless cited the issues of liability and indemnity provisions in both the TELUS and Bell tariffs as the primary reason for its unwillingness to offer Wireless E9-1-1 on a voluntary basis to its customers. In an attempt to bolster its case in a related setting, the carrier had made a submission to the CWTA working group and cited the U.S. Wireless Communications and Public Safety (9-1-1) Act to argue that wireless carriers should be entitled to better liability protection than that afforded in the

---

[6] This refers to the interconnection charge for Bell's 9-1-1 network access service. At the time, wireless carriers were offering only basic 9-1-1 service and thus being charged less than the CLECs that were offering full E9-1-1 service.

TELUS and Bell legal agreements developed in the CISC (Canadian Wireless Telecommunications Association, Wireless E9-1-1 Working Group, 2001b). This claim ignored the fact that in their defence, both Bell and TELUS had highlighted that their agreements drew directly upon a consensus document produced by members of the CISC Emergency Services Working Group for wireline E9-1-1 network access service to CLECs. Bell then seized an opportunity to criticize Rogers Wireless for its lack of participation in the CISC activities, observing that it 'has been largely absent from the [Emergency Services Working Group] and ... appears to have little or no understanding of the respective risks assumed by [Bell], [wireless carriers] and municipalities in relation to the delivery of 9-1-1 service [which] is telling' (Bell Canada, 2002, par. 26).

All this time, the PSAP representatives remained unusually absent from the comments on the Bell tariff notice. The Ontario E9-1-1 Advisory Board submitted no comments to the proceeding and not a single word was mentioned in conjunction with Bell's proposed Wireless E9-1-1 service offering with respect to the outstanding issue of entering wireless subscriber records into the ALI database. This odd circumstance may have had something to do with the fact that the ALI database issue had by that time reached an impasse in the CISC and the CRTC had a month previous to the Bell tariff filing issued Public Notice 2001-110 in an effort to break the stalemate. On 21 December 2001, the CRTC approved the Bell tariff on an interim basis, paving the way for the deployment of Wireless E9-1-1 in Central Canada—only the second region in Canada to make the service commercially available to wireless customers (Canadian Radio-television and Telecommunications Commission, 2001h). Microcell had pledged in its comments during the Bell tariff proceeding that it would seek to immediately enter into an agreement with Bell for Wireless E9-1-1 network access service and modify accordingly the Part VII application it had made requesting that the CRTC mandate the deployment of Wireless E9-1-1 network access services across Canada.

Hand-off to the CISC

Given the potentially complex interconnection issues stemming from paragraph 286 of Decision 97-8, the CRTC recognized that technical matters would need to be addressed in a setting that would ensure CLECs could provide E9-1-1 on a competitive basis to their customers. Over the course of a decade or more, the incumbent carriers had, during the monopoly-era interconnection regime, built and maintained 9-1-1 systems to serve major urban centres of Canada. With the advent of local competition, each of these 9-1-1 platforms now represented a potential bottleneck that the ILECs could use to their advantage. The CRTC understood that in order for the CLECs to meet the obligations of paragraph 286, there was a need for a corresponding service environment with which to ensure fair and equitable means of deploying or accessing facilities to support E9-1-1 service. The CRTC thereby directed its Interconnection Steering Committee (CISC) to fulfil this function by making 'recommendations concerning the appropriate arrangements for the provisioning of 9-1-1 service reflecting the competitive framework established in this Decision' (Canadian Radio-television and Telecommunications Commission, 1997, par. 287).

Following Decision 97-8, the CISC began work on the development of the 'Strawman'—a document representing high-level consensus among the ILECs and others on 9-1-1-related interconnection arrangements. This vehicle of inscription served as a basis of congruency among members and established a baseline for future negotiations between carriers. The Strawman is more formally titled the '9-1-1 Trunk-side CLEC Interconnection Document' and has since been adopted to serve as the template from which 9-1-1 interconnection arrangements are implemented in the first leg between ILECs and CLECs. It includes a broad range of considerations that cover the voice network, data transfer and management, service management and responsibilities, 9-1-1 service provisioning and network assurance (CRTC Interconnection Steering Committee, 2000). While it is clear that the CRTC in its local competition framework and the CISC in its various activities anticipated the advent of *fixed* wireless CLECs, it is far less apparent in reading the wording of the relevant documents that a *mobile* wireless carrier might one day apply for CLEC status. With the interim approval of the Microcell and Clearnet applications in September 2000, the CISC was directed by the CRTC to begin a process of reviewing and revising the Strawman, where necessary, to accommodate the unique aspects of mobile telephone service.[7]

Making Competition Work

The history of the CISC in fact predates Decision 97-8, beginning in August 1996 with Telecom Public Notice 96-28, 'Implementation of Regulatory Framework: Development of Carrier Interfaces and Other Procedures.' The CRTC had recognized prior to this proceeding the technical complexity inherent in the transition to a competitive telecommunications environment and, with specific reference to matters of unbundling and local number portability, moved to establish a committee to expedite decision-making and build consensus among industry stakeholders. Hence, the CRTC's Interconnection Steering Committee (CISC) was born, its first meeting held in September 1996. In most basic terms, the purpose of the CISC is 'to oversee the process of identifying the requirements and developing the administrative and support systems required to facilitate local telephone competition' (Canadian Radio-television and Telecommunications Commission, 1996b).

As noted above, the CISC can be seen as an institutionalized locus for reflexivity based on a public-participatory and consensus-driven mandate. I have also shown that the CISC process is different from strategic niche management to the extent that it acts as a space of legitimacy for interested parties to engage in

---

[7] There is a subtle but important distinction between wireless mobile and wireless fixed service. An example of wireless fixed service would be Local Multipoint Communication System (LMCS) that offers wireless local loop to a residence. This is a wireless service to a fixed address and does not offer the mobility that is made possible with a cellular network.

discussions about issues affecting network design, or more specifically, issues related to interconnection.

Activities within the CISC are either assigned by the CRTC or originated by the public and as such, fall within the Commission's jurisdiction.[8] Within the CISC, a steering committee is appointed to oversee and direct a number of Working Groups and 'Ad Hoc' Committees. Tasks related to Wireless E9-1-1 fall under the mandate of the Emergency Services Working Group (CISC/ESWG), which also deals with all matters pertaining to 9-1-1, including wireline services (See Figure 6.1).

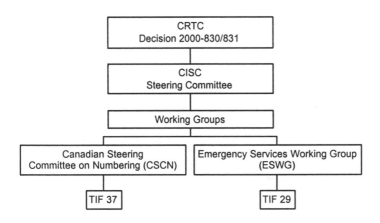

Figure 6.1 Wireless E9-1-1 within the CISC

Following the interim approval of Clearnet and Microcell applications, it was the CISC/ESWG that developed the Strawman and its various revisions to include Wireless CLECs. A number of other working groups are involved indirectly with Wireless E9-1-1, including the Canadian Steering Committee on Numbering (CSCN), which is concerned with the design of the Emergency Services Routing Digit system. The Network Security Working Group has also been implicated in Wireless E9-1-1 services (in conjunction with Microcell's allegations, noted above).[9]

The Emergency Services Working Group (ESWG) and others undertake their assigned duties as 'Tasks' that are introduced by a formal process and follow an

---

[8] My discussion of the CISC process and its organization is drawn from the CISC Administrative Guidelines (Canadian Radio-television and Telecommunications Commission, 2001d).

[9] Personal communication with the Chair of the Network Security Working Group revealed that the issue of the ALI database was indeed taken up but 'after brief discussion ... participants reached agreement that CISC may not be the best place to discuss those issues' [by email 9 May 2002]. CRTC Interconnection Steering Committee (1999) provides citation details for relevant documentation.

equally formal process while tasks are under consideration. A 'Task' may be initiated by the CRTC through the CISC Steering Committee or by a written proposal from any interested party directly to the Chair of a specific Working Group (Canadian Radio-television and Telecommunications Commission, 2001d, p. 4). Tasks proceed in three steps: initiation; work; and impasse or consensus. At the outset, a numbered Task Identification Form (TIF) is issued according to specific guidelines that require that the task is 'an identifiable deliverable that is expressed in an objective manner.' In effect, the TIF serves as a vehicle of inscription to bind parties into congruent action. Work on the TIF is done through Contributions made by parties 'to explain their views and propose alternatives for the completion of the task' (p. 5). Contributions from interested parties are key documents and provide much of the substance of a TIF, by setting out positions, establishing facts, and proposing solutions. The Working Group maintains an Activity Diary for each TIF, which includes minutes of each meeting, action items adopted, and the status of each action item. The desired outcome of a TIF is to issue a consensus report, such as the Strawman. The report goes through an approval process, then it is filed with the Chair of the Steering Committee, reviewed, and subject to CRTC approval, action deemed appropriate is taken (p. 6).

In the event that a dispute arises during the TIF proceeding, a Dispute Information Form (DIF) is prepared by the involved parties and reviewed by the Steering Committee to determine if further direction can be given to the Working Group in order that it may resume its activities. If this is not possible then the matter may be referred back to the CRTC. The Chair of the Working Group, however, is responsible for resolving issues before they escalate into disputes, and parties to the TIF are encouraged to 'refrain from developing entrenched positions, and ... [to] explore the viability of all reasonable options put forward at a meeting' (p. 6). Here we find an emphasis on learning within the CISC.

Most importantly, membership within the CISC and its Working Groups 'is open to all interested parties' and members share a responsibility to follow the rules of the proceedings, to be aware of issues, and to work toward solutions in an expeditious manner. Documents generated by the CISC and its Working Groups are considered part of a public process and are made available on the CISC website. In this respect the CISC can be taken as a public-participatory forum, however, certain limitations are recognized in the CISC Administrative Guidelines. According to its operating principles, for instance, the CISC is required to 'afford all parties the right to be heard on CISC-related matters' while 'recognizing that most CISC work is highly technical and therefore of limited interest to consumers.' As such, the CISC is therefore required to 'make reasonable efforts to inform consumer representatives of any consumer issues arising in the CISC,' although the terms and conditions of these efforts are not specified. Due to the expeditious nature of CISC activities, however, parties wishing to have their views heard on matters are expected to attend meetings and express those views 'in the course of work on the tasks.' Failing that, they 'should not expect to be able to have items re-examined' in the CISC process (p. 1). Given these constraints, the public nature of the CISC is perhaps more *pro forma* than *de facto* but it nevertheless could be open to a much wider constituency than technical working groups alone.[10]

The Limits of Reflexivity

Despite the mandate and formal procedures intended to reach consensus in an efficient manner, the membership of the CISC could not resolve the dispute over the ALI database and customer billing records. Following interim approval of Orders 2000-831 concerning the Microcell application for wireless CLEC status, the CRTC directed the CISC to address issues pertaining to interconnection arrangements. Members of the Emergency Services Working Group determined that the Strawman needed to be reviewed and revised and as a result, TIF 29 was initiated on 5 October 2000, with Bell Canada and Microcell issuing the first Contribution in the form of proposed changes to the Strawman (CRTC Interconnection Steering Committee, 2001). The Activity Diary for TIF 29 indicates that a total of ten contributions were made from the first entry in November 2000 to the last entry in August 2001.

Two months after the last entry into the Activity Diary, the CRTC decided to issue Public Notice 2001-110 to address the dispute that had been plaguing the Emergency Services Working Group (ESWG). Although the Working Group had been largely successful in revising the Strawman document and no formal disputes had been filed for TIF 29, the ILECS, wireless carriers, and PSAPs had reached a complete impasse over the design of the ALI database and the recalcitrant matter of populating it with wireless customer billing records. The controversial directive from Order 2000-831 could not be resolved. Although the wording of the CISC administrative guidelines encouraged parties to 'refrain from developing entrenched positions, and ... [to] explore the viability of all reasonable options put forward at a meeting,' the limits of reflexivity seemed to have been reached.

A Locus for Reflexivity

Earlier in the chapter I described 'loci for reflexivity' as an intervention strategy that attempts to expand the range of technology assessment discourse and thus it brings with it the prospect of new forms of boundary crossing and related challenges in achieving stakeholder consensus. Instances of boundary crossing within a technology project may lead to disputes because of perceived functional or territorial aspects of problem formulations or specific design proposition challenges. Yet often, as in data internetworking case, these disputes mask the more extensive normative views of interested parties, preventing a full grasp of the wider contextual concerns. Bijker's 'technological frame' and its associated configuration model provide a means to investigate instances of boundary crossing

---

[10] Criticism of the CISC's public process has come from Canada's Public Interest Advocacy Centre (Lawson, 1998).

in order to better understand how communities of experts interact within a technology project.

Boundary Crossing within Wireless E9-1-1

Boundary crossing is a useful concept with which to interpret the controversy and disputes surrounding the design and use of the ALI database in the Wireless E9-1-1 case in Canada. Several functional and territorial dimensions of boundary crossing are readily apparent in the case because Wireless E9-1-1 requires a number of significant changes in the telecommunications infrastructure, both at the core and the periphery, involving three major stakeholder groups (wireless carriers, ILECs, and PSAPs) and spanning the entire domain of interconnection space.[11] For all parties involved, Wireless E9-1-1 required changes in both equipment and operational procedures. It also presented an opportunity for Microcell to resolve the business problem of entering CLEC territory and addressing PSAP operational challenges, such as handling wireless emergency calls without the ANI/ALI function.

Let us take up the possibility that we have an instance here of boundary crossing, similar to Abbate's internetworking case, in which competing design propositions generate territorial conflict that is underpinned by competing normative assumptions. In other words, let us entertain the idea that we have here a configuration of one or more technological frames with respect to the wireless E911 project. Yet how do we establish the presence of one or more technological frame(s)?

It is clear that many of the elements of a technological frame were shared among all interested parties. Yet a number of issues indicate a competing technological frame, most notably those stemming from the disputed requirements for the ALI database. The public safety agencies persistently called for a design that included real-time access to customer billing records and introduced different elements to support such claims. The wireless industry largely regarded this call as illegitimate and adopted a dominant design proposition using emergency service routing digits (ESRD). A unified interpretation of one technological frame emerges, however, when we examine the ways in which all sides positioned themselves around the problem of providing access to customer billing records.

Recalling Bijker's configuration model, the presence of a single dominant technological frame tends to favour a conservative approach to problem formulation, with a division between high inclusion actors operating within the frame and low inclusion actors outside the frame. Bijker proposes that high inclusion actors will draw on standard problem-solving strategies based on an exemplary artifact while low inclusion actors may offer alternative approaches that disregard or challenge the exemplary artefact. While all parties agreed in principle

---

[11] Of course the general public is also a stakeholder. Given the limited representation of the public so far in the Wireless E9-1-1 proceedings, I am not comfortable discussing the public in this analysis. A separate study of public perceptions related to Wireless E9-1-1 would make a critically important contribution to policy research for this emerging service.

to the need for Wireless E9-1-1, clear differences were evident on the matter of access to customer billing records. These differences tended to form along the lines of wireless carriers and ILECs versus the PSAPs (with the later exception of the BC 9-1-1 Service Providers Association). My claim that a single dominant technological frame emerged is derived from the fact that all parties to the proceeding seem to have accepted a basic design proposition using Emergency Services Routing Digits (ESRD) and the existing ALI database system. This was evident in the strong support by all parties for the technical trials held in Alberta and Ontario. We can identify the ALI database as the 'exemplary artefact' at the heart of the Wireless E9-1-1 frame. For high inclusion actors such as the wireless carriers and ILECs, the ALI database, because it is designed with a single field for address information, is an obdurate, well-defined element with rigid parameters that present limitations on the design of the Wireless E9-1-1 service. As such, a decision was made that this address field would be populated with cell site location information to be triggered by an ESRD.

The PSAPs, however, would appear to be low inclusion actors on this matter. By low inclusion, I mean that they use the ALI database when handling emergency calls but otherwise have limited control over its design and implementation in the E9-1-1 system. Bijker uses the term 'boundary object' to describe the relationship of low inclusion actors to an exemplary artefact. As low inclusion actors, the PSAPs proposed that the ALI database *could* be used to provide real-time access to customer records, taking it to be a flexible artifact of opportunity, capable of delivering wireless customer billing records if necessary. One exchange between Microcell and the Alberta PSAP representative over the very existence of a set of technical standards offers a prime example of where this claim is disputed in distinctly functional terms but with clearly territorial overtones concerning control over access to customer information.

Does the dispute over the ALI database lead us back to the normative reference points on which the competing positions may be founded? The theory of a single dominant technological frame helps us to interpret the normative positions from which each side in the dispute applies the basic problem formulations for Wireless E9-1-1. For the wireless industry, the problem is formulated as a reverse salient. In other words, the wireless industry has sought backwards compatibility and ways to provide an E9-1-1 equivalent that will fit within established practices and cause the least amount of disruption (and cost). As a result, the design proposition for Wireless E9-1-1 in Canada is conservative and drawn wherever possible from methods developed in the United States that most closely conform to current capabilities and standard practice. As the prospective wireless CLEC put it, adding new functions to the ALI database as proposed by the public safety agencies would:

> greatly complicate the interaction between Microcell resellers and their end-user subscribers, cause Microcell and its resellers to incur substantial and unjustifiable ongoing compliance costs, raise important privacy concerns, aggravate and potentially drive away significant numbers of subscribers, and place Microcell at a distinct competitive disadvantage relative to other wireless carriers. All of this for a measure

whose rationale is dubious to begin with ... (Microcell Telecommunications Inc., 2001f, p. 5-6)

The public safety agencies, however, formulated the problem quite differently. In their view, dispatch operations had suffered at the hands of the wireless carriers because of the added workload when handling emergency calls made from mobile phones:

> ... we are tired of being the victims of the success of the wireless companies. Wireless calls have grown from 8 [per cent] of our total calls in 1990 to nearly 40 [per cent] in 2001. It takes extra time to route a wireless 9-1-1 call to the correct dispatch centre. It is becoming more difficult to accommodate this additional work without extra resources. (BC 9-1-1 Service Providers Association, 2001a)

Given the new demands placed upon them by the explosive growth of the wireless industry, the PSAPs saw it in their best interests to make demands for any and all information that they could locate in order to gain some control over the real-time flow of data. Certainly, this sense of urgency is to some extent the source of their demand for the inclusion of wireless customer billing records in the design of the Wireless E9-1-1 system.

Each problem formulation highlights different constellations of elements, giving credibility to some while ignoring others. At the heart of the territorial debate, however, is the ALI database and its status as a technical artifact. The PSAPs were low inclusion actors confronting a dominant technology frame, and from this perspective the ALI database was viewed as a boundary object that the wireless industry was using to prevent real-time delivery of wireless subscriber records. Attempts by the wireless industry to prevent subscriber records from populating the ALI database was interpreted by PSAPs as a territorial infringement on PSAP obligations to public safety operations.

The unfortunate result was an extended turf war over the capabilities of the ALI database perpetuated at the expense of wider contextual concerns that might result in alternative problem formulations. In the very late stages of the proceedings, for example, evidence was presented to point to further territorial elements at play in terms of the past history between the PSAPs and the wireless carriers. The proceedings eventually revealed that the wireless carriers' security departments had a poor record of responding to PSAPs requests for subscriber record information. The PSAPs accused the wireless industry of failing to take these requests seriously and provide adequate quality of service in dealing with requests from public safety agencies concerning 9-1-1 calls (BC 9-1-1 Service Providers Association, 2001a; Ontario 9-1-1 Advisory Board, 2001).

This revelation in fact did lead to an alternative design proposition for the problem of accessing customer records for investigative purposes. Introduced during the final submissions to Public Notice 2001-110, this design proposition was positioned well outside the technological frame that had dominated the previous proceedings because it did not involve the ALI database whatsoever. The proposal was to improve PSAP access to customer records by simply ensuring that the wireless carriers would provide improved quality of service in their security departments, which would mean easier telephone access to security staff and faster response to any inquiries made by a PSAP operator investigating an emergency call.

Had the interested parties not been so fixated on the ALI database and had the CRTC not fueled this debate with its obligation in Order 2000-831, the alternative problem formulation may have surfaced far earlier in the proceedings. A careful review of the proceedings reveals little apparent need for an extended debate about the merits and possibility of a design proposition based on real-time access to wireless customer billing records but rather, seems to suggest that the presence of a single dominant technological frame contributed to a fixation on the ALI database at the expense of recognizing an alternative that involved a far more simple and less controversial solution. Furthermore, had the regulator taken a more active role in assessing the contributions of the interested parties within the CISC and had the regulator taken greater initiative in making effective interrogatories during the prolonged debate over the ALI database, it is entirely conceivable that the simple solution finally arrived at might have surfaced without such prolonged and needless dispute.

Sequencing and Timing

In August 2003, the CRTC issued Decision 2003-53, putting an end to the difficulties that had started three years earlier with Order 2000-831 and culminated in Public Notice 2001-110. Among other things, Decision 2003-53 set out the CRTC's position on customer billing records and the ALI database, noting 'it would be more effective and cost-efficient for all parties to focus on improvements to the wireless emergency services and underlying network services' than an attempt to retain the controversial obligation for wireless CLECs in Order 2000-831. Wireless CLECs such as Microcell were still required, however, to provide wireless E9-1-1 where it was available from the ILEC and 9-1-1 services on a parity basis with the respective ILECs in all other regions:

> ... the Commission concludes that in a community where the ILEC provides basic 9-1-1 service, the wireless carrier must provide a comparable level of service, in order to attain or maintain status as a wireless CLEC. Similarly, in a community where the ILEC provides no 9-1-1 service, a wireless carrier may operate providing no such service (Canadian Radio-television and Telecommunications Commission, 2003).

Perhaps more significantly, the CRTC extended the E9-1-1 obligation to all wireless carriers, thereby relinquishing its policy of regulatory forbearance in the interest of public safety and in light of the growing impact of mobile phones on the public safety agencies. The CRTC did not comment, however, on its previous decision in the Local Competition framework to deny 9-1-1 'essential service' status, which seems to indicate that the terms and conditions of interconnection in the *de facto* bottleneck of 9-1-1 platforms in most parts of Canada remains at the discretion of the incumbent carriers.

The dispute over customer billing records and the ALI database was solved with the CRTC's acceptance of an alternative proposition:

> ... it was suggested that a wireless CLEC should be required to staff its operations centre on a continuous basis (i.e., 24 hours per day, seven days per week) so as to be able to respond to queries for subscriber information by authorized PSAP personnel. It was also suggested that wireless CLECs should be required to maintain a toll-free number for this purpose.

> In the Commissions view, this type of mechanism would be a cost-effective way of improving emergency services since it would augment the information available through wireless E9-1-1 service capabilities. The Commission therefore requires wireless CLECs to establish and maintain ... toll-free telephone access to and continuous staffing of at least one of their operations centres, in order to promptly assist authorized PSAP personnel seeking subscriber information in emergency situations. (Canadian Radio-television and Telecommunications Commission, 2003)

Again relinquishing its policy on regulatory forbearance and in conjunction with the related E9-1-1 directive, the CRTC also extended this obligation to all wireless carriers. Yet, why were these seemingly simple solutions to the ALI database issue not resolved within the CISC? After all, according to its own mandate and guidelines was it not the appropriate forum in which these parties were supposed to exchange views and formulate alternative design propositions on matters of interconnection and network design? However effective and efficient the CISC had been in terms of working out network access services and other lower layer interconnection matters, in this case it clearly revealed some limitations as a locus for reflexivity.

What the case suggests more generally is that in any given technology project there are likely to be numerous spaces of legitimacy for interested parties to exchange views and yet not all spaces are appropriate for all discussions. The CISC, it appears, is a forum best suited to well-framed technical matters of business—established and approved design propositions, as it were. As such, the CISC is not likely to be as effective when dealing with alternative problem formulations and design propositions offered by 'low inclusion' actors such as the PSAPs. While the PSAPs are hardly minor stakeholders in Wireless E9-1-1, they nevertheless appear to have found themselves outside the dominant technological frame that presided over the CISC Emergency Services Working Group. As a result, the design discourse was continuously revolving around an exemplary artifact—the ALI database—and its capabilities and limitations. Behind this apparently functional dispute was a territorial issue stemming from the frustration of the PSAPs in gaining access to customer information for their follow-up investigations. The poor performance of the wireless carriers in responding to previous PSAP requests for customer information was regarded by the PSAPs as an infringement on the PSAP obligations to serve the public. In the form of a public hearing, Public Notice 2001-110 provided the locus for reflexivity that permitted a wider discourse on the matter, abolishing some of the dominant technological framing that had plagued the proceedings. The public hearing enabled an alternative problem formulation to surface by acknowledging the legitimacy of PSAP concerns outside the scope of debate that had so tightly restricted the dispute up to that point.

Finally, the Wireless E9-1-1 case suggests that the framing of disputes within a technology project may conceal the normative foundations on which such disputes are based, suggesting that an intervention strategy that seeks to bring together communities of experts through loci for reflexivity must also account for divergent *cultures of expertise*. Cultures of expertise bring with them a range of normative perspectives that are essential to reducing risk and vulnerability in the management of critical infrastructure, however, not all cultures of expertise are necessarily appropriate in all settings and at all points in time. As such, attempts to establish effective spaces of legitimacy requires an element of appropriate sequencing and timing, as will be described in the concluding chapter.

Chapter 7

The Structures of Intervention

Wireless E9-1-1 has now been deployed across much of Canada, with the CWTA and others currently involved in discussions about implementing a high-resolution version similar to the FCC's Phase Two requirement. It remains to be seen if the interested parties will experience similar successes and challenges and whether the CRTC will take a more proactive role in this latest technology project in public safety telecommunications. Given the results of the study on the first phase deployment of Wireless E9-1-1, what kind of advice might we offer to those interested parties now embarking on these phase two proceedings? Or more generally, what can we learn from the case study taken up in this book to inform a wider field of policymaking for critical infrastructure?

My central argument in this book is that a program of mitigation-oriented policy research must expand its agenda to focus on the social roots of risk and vulnerability. When applied to the management of critical infrastructure, this research agenda suggests two primary objectives: first, to understand the process of growth and change in large technical systems; and second, to identify and assess intervention strategies that are capable of reconciling wider public interest with the exigencies of a mitigation strategy. The intent of this book has been to establish a theoretical and conceptual foundation for the claim I present and to introduce an empirical approach capable of supporting a program of research consistent with this claim. The case study has been conducted in an effort to demonstrate that such research can be done and that it can achieve policy relevant results. My aim in this chapter is to delineate that 'last mile' between the findings of the specific case study and some of the implications it suggests for the management of critical infrastructure more generally.

Whereas the previous three chapters report on a 'sociological deconstruction' of Wireless E9-1-1 in Canada, I begin here with a brief summary and socio-technical mapping of the case before moving to an assessment of the layout of interventions evident in the case. This succession follows the steps necessary to form a constructivist study of contemporary technology projects, as described in chapter two (see Table 2.1). These preliminary steps are intended not only to provide a summary of the case but perhaps more importantly, to illustrate the application of the intervention matrix as an analytical tool for the assessment of interventions. The summary observations and the policy recommendations derived from them were direct results of the application of the intervention matrix to the layout of interventions found in the case. I have structured this concluding chapter with two objectives in mind. On the one hand, I wish to illustrate that the methodology developed in this book can be used to support policy research efforts.

On the other hand, I wish to draw on the applied methodology to provide some insight for policymakers who must contend with the management of critical infrastructure among their portfolio of responsibilities, particularly mitigation-oriented initiatives of one kind or another.

Influencing Growth and Change

We begin with a socio-technical mapping based on a summary of the identified actors and issues in the Wireless E9-1-1 case study. Arnbak's functional systems model serves as the groundwork for this mapping, making it clear that all dimensions of interconnection space have been important to the development and early deployment of Wireless E9-1-1, with each layer tending to circumscribe a specific set of actors and issues. Table 7.1 charts the case overall, listing key factors in each of the functional layers of interconnection space and the associated issues.

Table 7.1 A socio-technical mapping of Wireless E9-1-1

Functional Layer	Factors	Issues
Information Services	WSP customer-activation process; subscriber billing record; MSAG.	Disruption of business models; control over critical information; customer privacy rights.
Value-added Services	ALI database.	Terms and conditions of access; database design and alternatives.
Network Services	Network access services (Strawman; E911 tariff); ILEC/PSAP contracts.	Terms and conditions of interconnection, including liability; requirements for new equipment; stranded investments.
Physical Infrastructure	Cellular radiotelephone networks; trunking options; locating cell-sites; province-wide E911 platforms and ILEC switches.	Problem equipment (one-button phones); legacy platforms; regional differences in network equipment and design.

At each layer of interconnection space one or more actors and issues had an important influence in this particular technology project. For instance, at the physical infrastructure layer, the growth of cellular radiotelephone networks undermined the original E9-1-1 design concept, creating a problem in locating mobile phone customers. Trunking provisions and signalling, switches, and cell-site mapping were all key factors at this lower layer, partly because of regional differences among the incumbent wireline carriers (ILECs). At the network services layer, there was a requirement to develop a consensus (the so-called

'Strawman') document to specify interconnection arrangements between wireless carriers and the incumbent carriers operating the 9-1-1 tandems. There were also additional contracts specifying network services and administration details between wireless carriers, incumbents, and the public safety answering points.

In the upper layers, the ALI database was a value-added network element operated by the incumbents to provide location information from the wireless carriers to the public safety agencies. At the functional layer of information services, the issues of customer billing records and access to the Master Street Address Guide (MSAG) featured prominently in a number of problem formulations and design propositions. The socio-technical mapping illustrates the heterogeneity of actors and issues involved in the case, showing how configurations may vary across interconnection space. This mapping helps to explain the multiple intervention strategies that were also evident in the case and as I shall point out, provides some important considerations for the assessing the layout of interventions.

The Layout of Interventions

Seven predominant intervention strategies are apparent in the development and deployment of Wireless E9-1-1 in Canada. Paramount among these is probably the CRTC's local competition framework based on Decision 97-8, which acts as a form of 'soft' technology forcing. More specifically, soft technology forcing is found in the principle of regulatory symmetry embodied in paragraph 286 of Decision 97-8, as it requires all regulated carriers to provide enhanced 9-1-1 service functionally equivalent to that of the local ILEC. This requirement is a demand-side strategy that forced Microcell in its application for wireless CLEC status to upgrade its network to offer Wireless E9-1-1 service in the first place. When faced with the problem that such an upgrade was not likely to be possible in all parts of Canada in the near future, the CRTC directed Microcell to populate the local ALI database with its customer billing records in locations where E9-1-1 service was not offered by the ILEC. In some respects this directive in Order 2000-831 was simply an extension (albeit a controversial one) of the soft forcing strategy embodied in Decision 97-8. Conversely, Microcell's Part VII Application to the CRTC was also a technology forcing effort, for it asked the regulatory agency to demand that the incumbent carriers commit to a specific upgrade path in network access services in order to enable future deployment of Wireless E9-1-1.

In terms of active Wireless E9-1-1 service, its development was carried out with the voluntary cooperation of the wireless industry. The Canadian Wireless Telecommunications Association (CWTA) adopted a form of strategic niche management as a way to develop supply-side initiative in order to acquire a technical capability and to undertake testing of Wireless E9-1-1 under the auspices of a specific working group. Reflexivity strategies that modulated between demand-side and supply-side efforts were evident in the CISC working group that sought, under specific direction from the CRTC, to establish consensus between the telecom sector and the public safety agencies on various terms and conditions of network access for Wireless E9-1-1. CRTC Public Notice 2001-110 could also be viewed as a strategy of reflexivity to the extent that it sought to clarify, among a range of interested parties, certain outstanding disputes and questions related to Wireless E9-1-1 in Canada. Table 7.2 summarizes the various intervention strategies evident in the Wireless E9-1-1 case.

Table 7.2 Layout of interventions in the Wireless E9-1-1 case

Actor/event	Strategy	Objective
CRTC Decision 97-8: Local Competition Framework	Technology forcing	E911 obligation forces symmetry on regulated carriers.
CWTA/Wireless E911 Working Group (WEWG)	Strategic Niche Management	Supply-side trials and system testing.
CRTC Order 2000-831	Technology Forcing	Resolve symmetry gap between 97-8 requirement and wireless CLECs.
Microcell Part VII Application	Technology forcing	Require ILEC network upgrades to support Wireless E911.
CISC/Emergency Services Working Group (ESWG)/ TIF 29	Reflexivity	Establish consensus on terms and conditions of network access services for Wireless E911.
CRTC Public Notice 2001-110	Reflexivity	Clarify outstanding disputes related to Wireless E911.
CRTC Decision 2003-53	Technology Forcing	Require all wireless carriers to implement Wireless E911 and related obligations.

Inscription and Action

Each intervention strategy in the development of Wireless E9-1-1 is based on a vehicle of inscription that provides the principal alignment around which the actions of affected parties are influenced. I proposed in chapter two that a strategy based on technology forcing relies predominantly on a vehicle of inscription. In the case of Wireless E9-1-1, the vehicles of inscription that support forcing strategies are found in CRTC Decision 97-8 (paragraph 286) and the formal Rules of Procedure, wherein the requirements for filing Part VII applications and governance on interaction between the regulatory agency and other stakeholders are both set out. The terms and conditions of these vehicles of inscription were established prior to and irrespective of Wireless E9-1-1 and for the most part are largely procedural in nature. In other words, while they were operative prior to this particular technology project, they are not intended to address directly the specifics of individual cases but rather to provide general policy directives and guidance. Table 7.3 summarizes the major vehicles of inscription identified in the case study.

Table 7.3 Vehicles of inscription and Wireless E9-1-1

Strategy	Actor	Inscription
Forcing	CRTC	Decision 97-8 (par. 286); Order 2000-831
Strategic Niche Management	CWTA	Wireless E911 Working Group (MOU)
Forcing (attempt)	Microcell through CRTC	CRTC Rules of Procedure
Reflexivity	CISC	Emergency Services Working Group (TIF 29)
Reflexivity	CRTC	Public Notice 2001-110
Forcing	CRTC	Decision 2003-53

As for the CWTA Wireless E9-1-1 Working Group and its voluntary efforts, a Memorandum of Understanding (MOU) between participating organizations provided the necessary vehicle of inscription for the technical trials to take place. In contrast to paragraph 286 of Decision 97-8 or the CRTC's Rules of Procedure, however, the MOU document was a limited term, project-oriented agreement that included detailed specifications on technical and administrative matters. It did not attempt to establish ongoing commitments or agreements between parties with respect to Wireless E9-1-1, but rather, it served to establish a basic design proposition from which interested parties could work together to undertake experimental testing.

In the strategy of consensus building central to the mandate of the CRTC's Interconnection Steering Committee (CISC) Emergency Services Working Group (ESWG), the vehicle of inscription was held/found in a task identification form, known within the working group as 'TIF 29,' which set out the terms and conditions for establishing network access services needed to provide Wireless E9-1-1. Like the CWTA Memorandum of Understanding, this vehicle of inscription was task-oriented and represented a certain acceptance of the dominant technological framing of the Wireless E9-1-1 project, as its main purpose was to establish a basis for specific negotiations toward the creation of a revised template or 'Strawman,' document. The revised Strawman has since become a high-level consensus document from which specific terms and conditions of implementation can be negotiated between parties implementing technical details of Wireless E9-1-1 service.

The reflexivity strategy evident in the CRTC public hearing under Public Notice 2001-110 was derived from the controversial obligations contained in the previous Order 2000-831, concerning Microcell's obligations as a wireless CLEC.

It should be noted, however, that the issues presented in the public hearing did not call directly into question the E9-1-1 obligation within the Local Competition Framework but instead solicited responses from interested parties on the specific issues and disputes stemming from Order 2000-831. For its part, Order 2000-831 was a response to Microcell's wireless CLEC application and the 'symmetry gap' it created in areas where Wireless E9-1-1 was not available. As such, it was a ruling on a specific case and was therefore limited in scope as a general effort at technology forcing. Had the Order stood, it could have been precedent-setting for other wireless carriers seeking CLEC status and acquired a more general technology forcing effect. By contrast, Decision 2003-53 contains the directives to establish a regulatory intervention based on technology forcing that requiring *all* wireless carriers to provide measures related to public safety obligations and Wireless E9-1-1 wherever it is available.

Establishing Congruency

I proposed in chapter two that effective strategic niche management is dependent on congruency between interested parties. Congruency refers to a shared perspective (not necessarily consensus) between stakeholders within a technology project. A shared perspective may be based on a common technological frame, although it may also be guided by other orientations such as public safety or emerging market opportunities (e.g., location based services). In the case of Wireless E9-1-1, the interested parties within the CWTA established a limited degree of congruency around the need for technical trials (see Table 7.4). This limitation was reflected in the specific of the CWTA Wireless E9-1-1 Working Group mandate, and evident in the sharp decline of the group's activities following the initial success of the Alberta and Ontario trials, even though other issues remained to be resolved before the service could be fully deployed. This observation, combined with the marked dispute that later emerged over customer billing records, suggests that once the trials were completed, a significant measure of congruency dissolved as parties moved to develop formal arrangements through the CRTC's Interconnection Steering Committee.

Strategies of technology forcing and reflexivity are also built on some degree of congruency, even though it is not the pivotal factor on which they rely. For example, the CRTC's initial role in Wireless E9-1-1 was based on a principle of regulatory symmetry to maintain fair competition, expressed in Decision 97-8 by establishing a distinction between regulated service providers and the forborne wireless carriers. Microcell's Part VII Application was initiated on the shared principle of congruency among stakeholders in the telecom sector, who recognized and responded to requests for cooperation provided that the intervening parties complied with the CRTC's Rules of Procedure. For instance, while not all interested parties accepted Microcell's specific claims to relief, all parties demonstrated congruent orientation toward the principles of the Part VII Application proceedings by their recognition of and compliance with the CRTC's interrogatories.

Table 7.4 Congruency and Wireless E9-1-1

Strategy	Actor	Congruency
Forcing	CRTC (Decision 97-8) and (Order 2000-831)	Fair competition through regulatory symmetry
Strategic niche management	CWTA Wireless E911 Working Group	Need for technical trial
Forcing	Microcell (Part VII application)	Accepted procedures for seeking CRTC intervention
Reflexivity	CISC Emergency Services Working Group	Negotiated rule making (CISC admin. guidelines)
Reflexivity	CRTC (PN 2001-110)	Public hearing and dispute resolution
Forcing	CRTC (Decision 2003-53)	Public safety

The principle of congruency for Public Notice 2001-110 was established according to the CRTC's right to revisit past decisions and to order a public hearing to resolve the disputes and other questions that had halted progress in the CISC. Congruency in the case of Decision 2003-53 is based on public interest obligations stemming from the widespread diffusion of mobile telephones. It is not clear, however, that congruency is well established with this Decision because it fails to address an inconsistency between the status of 9-1-1 as a non-essential service in the Local Competition Framework and the forbearance directive in the *Telecommunications Act*.[1]

Legitimacy and Participation

The development of Wireless E9-1-1 in Canada involved a number of spaces of legitimacy that influenced to some degree the participation of certain stakeholders. I proposed in chapter two that a strategy based on loci for reflexivity relies on an appropriate space of legitimacy, and two instances of such a strategy are evident in the CISC and CRTC, both providing opportunities for a wide range of interested parties to participate in a technology project. The CISC Administrative Guidelines establish the CISC as a forum for negotiated rule-making open to all interested parties despite the highly technical nature of its discussions. The CRTC Public

---

[1] The *Telecommunications Act* encourages the CRTC to forbear from regulating competitive (or potentially competitive) services. 9-1-1 service, because it was not classified as an 'essential service' in Decision 97-8, therefore falls into the category of a competitive or potentially competitive service, something that Clearnet PCS had indicated to the CWTA working group. Wireless carriers not seeking CLEC status could draw on this inconsistency in the local competition framework to challenge the CRTC's directive in 2003-53 that requires them to provide Wireless E911 service.

Notice process is also open to participation by all interested parties. Despite this, the levels of involvement and reflexivity in these processes were constrained by other factors, partly related to congruency.

In the other instances of intervention, stakeholder legitimacy was based on membership or invitation, as with the CWTA Wireless E9-1-1 Working Group or by a specific request for information based on the interrogatory procedures established by the CRTC for handling Part VII applications. Table 7.5 provides a summary of stakeholder legitimacy and participation in the development of Wireless E9-1-1 in Canada.

Table 7.5 Legitimacy of participation in Wireless E9-1-1

Strategy	Space/process of legitimacy	Basis for legitimate participation
Forcing	CRTC (Decision 97-8) and (Order 2000-831)	Public process leading up to Decision and Order
Strategic niche management	CWTA Wireless E911 Working Group	CWTA discretion
Forcing	Microcell (Part VII application)	Interrogatory from CRTC
Reflexivity	CISC Emergency Services Working Group	'Public' forum for negotiated rule-making
Reflexivity	CRTC (PN 2001-110)	Public process
Forcing	CRTC (Decision 2003-53)	Public process leading up to Decision

Observations and Patterns

Having identified the predominant intervention strategies in the case study, the next step is to set the strategies against the interconnection matrix and then look for larger patterns in the overall development of the case. The intervention matrix developed in chapters two and three combines the three conditions for successful stakeholder alignment with CTA's intervention strategies and Arnbak's functional systems model to create a three-dimensional framework for conducting systematic analyses of technology projects.

One important pattern to notice from the case is that a unique functional layer in interconnection space is predominant in each observed intervention strategy (see Table 7.6).

In the case of the CRTC's Local Competition Framework established in Decision 97-8, E9-1-1 obligations are prescribed in terms of functional requirements for a value-added service. Paragraph 286 simply directs CLECs to provide symmetrical information content to the ILEC in the form of an E9-1-1 application. Such a definition positions the directive in the upper layers of interconnection space where applications and information content are defined, but there is little regard for the details of technical implementation. The directive is intended to be a technology-neutral forcing strategy based on a kind of functional equivalency. It makes no provisions for the involvement of stakeholders or the consultative processes in the actual design and deployment of equivalent services, such as Wireless E9-1-1, other than to direct the CISC to the details of interconnection.

Table 7.6 Comparing Wireless E9-1-1 with the Intervention Matrix

	Forcing	Niche Management	Reflexivity
Legitimacy	CRTC	CWTA WEWG	CISC ESWG
Congruency	Fair competition	CWTA technical trials	Consensus
Inscription	Decision 97-8 (par. 286); (Order 2000-831)	MOU	TIF 29
Interconnection Space	Upper layers	Coordinate upper layers with lower layers	Lower layers

In contrast to the functional regulatory directives, the CISC deals primarily with the very technical details of interconnection between network services and the lower layers of network access services and physical transport. The CISC Emergency Services Working Group (ESWG) operates under the auspices of the CISC, which as we have seen, is intended to be open to all interested parties as indicated in its statement of Operating Principles:

> The CISC shall ... afford all parties the right to be heard ... support the evaluation and acceptance of issues and development of resolutions based on their merit ... make reasonable efforts to inform consumer representatives of any consumer issues arising in the CISC ...

Yet, the task-oriented and consensus-driven mandate of CISC also limits the scope of discussion that can be undertaken in this locus of reflexivity:

> ... Recognize that broad and consistent achievement of a consensus resolution is a fundamental expectation and the reason for the existence of the CISC ... most CISC work is highly technical and therefore of limited interest to consumers ... (Canadian Radio-television and Telecommunications Commission, 2001d, Sect. 3)

In the case of Wireless E9-1-1 design, the late stage of CISC ESWG involvement placed practical limitations on the range of alternative design propositions that could be tabled for discussion. In effect, TIF 29 was a second-order undertaking based on a first-order assumption—namely, that consensus on a suitable design proposition for Wireless E9-1-1 had already been achieved. The task for the CISC ESWG, therefore was not to debate the merits of the design proposition *per se* but rather to negotiate terms and conditions for network access services, despite the fact that certain participants continued to regard this design proposition as unacceptable. This observation implies, perhaps ironically, that the CRTC forum to evaluate and implement technical designs based on 'merit' is perhaps invoked too late in the process, when the momentum of a technology project is such that alternative design propositions are more or less dismissible. This would suggest that the role of the CISC in the design nexus is not well suited for a debate on alternative concepts for upper layer value-added and information services. Rather, a more appropriate role for the CISC is to implement accepted design propositions by establishing terms and conditions of lower layer interconnection arrangements.

Given this potential limitation of the CISC, one might reconsider the view that CRTC public hearings are therefore the appropriate spaces of legitimacy for a wide discussion of issues at higher functional layers. While regulatory hearings may provide an open forum for debate on alternative higher layer design propositions, these opportunities are constrained by two factors: the regulator must see fit to conduct a hearing on an issue; or an interested party must file a specific request for relief based on the CRTC Rules of Procedure (which could in turn lead to a public hearing). These constraints place a real limit on the ability to introduce new items to the regulatory agenda and may in fact restrict public input on issues in the higher layers of interconnection space. The proceedings leading to the CRTC Local Competition Framework issued in 1997 presented the last occasion for open debate on E9-1-1 obligation in Canada, and the record indicates that such a debate *did* take place (Canadian Radio-television and Telecommunications Commission, 1997, par. 285). Once the CRTC issued its policy framework, however, the regulatory status of E9-1-1 was effectively established, or 'black boxed' to use a term borrowed from the constructivist literature. This closure then restricted the range of future debate that would be acceptable within a CRTC public hearing on Wireless E9-1-1. For instance, neither Public Notice 2001-110 nor the subsequent Decision 2003-53 subjects the E9-1-1 obligation to fundamental debate. Rather than opening the black box and taking a closer look at the E9-1-1 requirement and in particular its status as a non-essential service, the public notice confined public input to specific issues and concerns arising from directives to Microcell in Order 2000-831. While the public hearing process may offer a space of legitimacy for public input, it may be an impractical locus in which to debate new service concepts or to assess alternative design propositions that challenge established definitions or accepted practices.

Where such debates might have taken place is within industry itself, where innovation and experimentation could have been regarded as potential opportunities for new markets in location-based services. Yet, in the case of the CWTA Wireless E9-1-1 Working Group, participation was limited to specific stakeholders invited into the process. Moreover, discussions were about learning only insofar as they established the conditions necessary to undertake technical trials. Further, as I have indicated in previous chapters, other issues and alternative design propositions were barely discussed within the CWTA working group, which tended to take a rather conservative approach to problem formulation on the matter.

Sequencing and Timing

An analysis using the intervention matrix reveals an overall pattern of significant constraints on experimentation and innovation when Wireless E9-1-1 was being developed and deployed in Canada. In effect, the analysis suggests that there is a fundamental drawback in the structure of intervention evident in the case of Wireless E9-1-1. Where the strategy centred on a locus for reflexivity, intervention was directed at the lower layers of interconnection space where participation was effectively limited by the highly technical nature of concerns, such as those addressed in the CISC. Where the strategy centred on technology forcing, intervention was directed at the higher layers of interconnection space, according to the Decision 97-8 technology-neutral specification for the functional requirements of E9-1-1 service.

An alternative structure of intervention would be one in which the strategy of reflexivity promotes discussion at the higher layers of interconnection space, where conceptual freedom allows for greater exploration of ideas and interplay among a diversity of stakeholders. Similarly, the strategy of technology forcing is more appropriately situated in the lower layers of interconnection space, where performance-based criteria can be closely assessed against current and future technical capabilities. Strategic niche management resides in the crucial middle ground, providing an intermediary step between experimental visions and the enabling of such visions through appropriate interconnection arrangements.

A clearly defined procedural ordering—or sequencing—in the flow of consultation and decision-making could improve the coordination of stakeholder groups with different ranges of expertise and experience, thereby reducing the risk of problematic patch dependency and encouraging innovative thinking. The overall assessment of interventions in the case of Wireless E9-1-1 in Canada suggests that sequencing and timing are essential considerations if policymakers are to effectively influence long-range management of critical infrastructure that is consonant with the principles of a mitigation-oriented policy framework.

Intervention Structures

The section to follow uses the analytic framework developed for this study to establish a set of recommendations for policymakers and practitioners involved in the management of critical infrastructure. The recommendations, based on three principles advocated throughout this study and applied to the analysis of a new service development in Canada, are as follows:

- Maximize learning and reflexivity through widest possible stakeholder participation in the management of critical infrastructure.
- Encourage innovation and experimentation on a broad scale.
- Ensure a level playing field and fair opportunities for new contributions in value-added services.

These principles were informed by a set of questions that were introduced in chapter three and then applied to the case study on public safety telecommunications. The first among these is related to the public policy objective to maintain 'orderly development' of the public information infrastructure and the corresponding claim that the infrastructure should be capable of supporting the broader objectives of a national mitigation strategy to 'safeguard' the social and economic fabric of a country. In the face of rapid technological change and regulatory reform, the issue raises the question of where and when direct supervisory or regulatory action may be required to ensure that critical infrastructure will evolve in a coordinated way and according to established best practices that embody a vision of long-term risk reduction. Accordingly, the issue leads to other key questions: When and where is supervision or regulation necessary? Is supervision or regulation currently in place? If so, can it be improved? If not, what are the best means to implement it given the constraints in the political and regulatory context?

Regulatory reform in Canada and in many other countries has resulted in requirements for increased reliance on market forces for service provision, meaning that many initiatives taken up under a mitigation policy framework should be fostered by competition wherever feasible. In terms of critical infrastructure management, this means identifying appropriate policy instruments to promote research and development and encourage technology transfer for business continuity planning, emergency management, or other related applications. It also will likely mean that regulators will need to ensure access to certain strategic network elements and services that enable experimentation and innovation, and this consideration introduces another set of questions. What policy instruments are available to support research, development, and technology transfer for disaster management? What are the key bottlenecks and strategic network elements that might influence the deployment of new services?

Public policy in many countries directs regulatory agencies to ensure that the social and economic requirements of user communities are considered in the evolution of critical infrastructures. In terms of mitigation-oriented policy initiatives, this suggests a need to ensure a diversity of stakeholder consultation in the management of critical infrastructure. Current stakeholder involvement may need to be expanded in new directions, or new processes may need to be introduced in order to better facilitate participation in 'keystone' technology projects. These considerations raise the following question: What are the important issues, challenges, and opportunities for expanding stakeholder participation in the management of critical infrastructure?

Learning and the Honest Broker

A key finding from the study was that some regions in Canada are currently unable to provide the minimum infrastructure required for the deployment of advanced value-added telecom services. In effect, this creates a patchwork of capability across the country, generally following and perhaps reinforcing patterns of economic development. For example, evidence from Microcell's Part VII Application to the CRTC indicates that Canada's public information infrastructure does not conform to a single standard of performance capability. On the contrary, this capability varies widely by region and the capacity for deploying new technology and services may depend on certain provisions at the lower layers of interconnection space such as switch platform upgrades and other physical and network access service support features currently not uniformly available across Canada, particularly outside the major urban centres.

If critical infrastructure is to provide a robust service environment to support mitigation-oriented policy initiatives, it will be necessary to ensure regional capability to meet specified minimum standards of infrastructure performance within a reasonable timeframe. In some cases, specified minimum standards may not have been established for matters involving emerging services and certain elements of critical infrastructure systems. Evidence from the case study also indicates that keystone standardization initiatives are likely to be influenced by a small number of dominant organizations, as was the case with Wireless E9-1-1 technical standards, which may limit direct participation in these undertakings.

In other instances, this situation of limited participation may precipitate misunderstandings and false expectations among certain stakeholder groups and lead to calls for 'gold-plated' standardization initiatives within otherwise ill-suited regional contexts, perhaps resulting in disputes and delays in the deployment of new services or eventual and costly path dependency problems. For example, in the case of Wireless E9-1-1 in Canada, some of the misunderstandings and enmity between parties might have been avoided had a third-party broker liaised between standardization activities with major organizations in the United States such as the TIA and NENA and the regionally specific interests, aspirations, and constraints of stakeholders in Canada. Such a broker might have been charged with vetting detailed technical information and closely assessing technical arguments, thereby providing a balanced perspective and accurate information to all parties including the regulatory agency itself. A broker might have been able to improve the efficiency of the overall process and, moreover, identify and encourage the adoption of verifiable best practices with an eye to critical path dependency issues (such as transitioning to FCC Phase Two-equivalent deployment or extending E9-1-1 to VoIP services). In the long run, this approach might have saved time and money for all parties involved (including customers and tax payers), while ensuring the orderly management of the public information infrastructure.

The honest broker function is an appropriate role for direct government supervision in response to the public policy objective of ensuring the orderly development of critical infrastructure. It should be the responsibility of national or regional governments *to facilitate* the identification, dissemination, and assessment of best practices in products and services from around the world, including the technical standards that contribute to the public policy objective of ensured orderly development of long-term investments in critical infrastructure. This responsibility does not mean that government should necessarily become directly involved in vetting details of the technical merit of such products and services, but rather, government should perform the role of honest broker to ensure the widest possible range and fairest terms and conditions of debate among interested parties. A relevant government ministry or department could, for example, sponsor a regular program open to interested parties with a clear intent to foster learning and elaboration on potential opportunities through dialogue, debate, and information sharing. Outcomes of such a program could clarify disputes or misunderstandings among users groups, generate new opportunities for innovation within the critical infrastructure sectors, and alert regulatory agencies to future items of interest or concern. Such a program would be an intervention strategy based on a locus of reflexivity with low barriers to entry and participation but clearly established working methods to ensure timely and effective exchanges and outcomes.

An exemplar for the design of such a program is found in the international domain with Project MESA. Project MESA (Mobile Broadband for Emergency and Safety Applications) is a technology standardization forum sponsored by the U.S. Telecommunications Industry Association (TIA) and the European Telecommunications Standards Institute (ETSI). First established in 2000 as the Public Safety Partnership Program (PSPP), it was renamed 'MESA' in 2001 although its primary aim remained the same: to create specifications for 'an advanced digital mobile broadband standard' to serve the public safety and emergency management communities (Project MESA, 2003). One important feature of Project MESA is that it seeks to promote constructive participation with the user community at the initial stages of technology design. Here it shares features with Constructive Technology Assessment:

> Within Project MESA, market relevance is achieved by inviting users and their organizations to play a focal role in the specification and harmonization of requirements *before the technical specification efforts are introduced.* MESA represents the first such international initiative within the ICT sector to 'put the user in the driver seat.' [emphasis added] (Project MESA, 2001)

The project has been designed to proceed in two general phases. The first phase involves the elaboration of a common statement of requirements (SoR) that covers 'a harmonized view of applications and services' for public safety and emergency management. The second phase involves the development of technical specifications in response to the common statement of requirements. The transition from the first phase to the second is intended to provide a preliminary set of specifications for service design, with implications for interconnection at both lower layer physical elements and network protocols as well as upper layer system and service aspects. The anticipatory nature of Project MESA is emphasized in the introduction to its SoR documentation:

The users of professional wireless telecommunications equipment within the sector of Public Protection and Disaster Relief (PPDR) have developed the MESA Statement of Requirements (SoR) document. It describes the services and applications, which a future advanced wireless telecommunications system should be able to support in order to realize the most effective operational environment for the sector.

Emphasis has been placed on those applications, which current applied technology cannot carry out to the full, but which have been identified by the users and their agencies to be key requirements.

The [SoR] document is unique in the sense that it represents the first trans-Atlantic consolidated view expressed directly by the professional users of advanced wireless telecommunication equipment. (Project MESA, 2002)

Applications identified in Project MESA include remote patient monitoring, mobile robotics, and ad hoc networking. To facilitate user participation, MESA is organized into a service specification group (SSG) and a technical specification group, both acting under guidance of a steering committee. The SSG deals specifically with issues relevant to the user community such as incident scenarios, quality of service concerns, and specific user requirements. Results from the SSG consultations are provided to the technical specification group to work out the technical matters entailed in transforming these service concepts into functioning systems. Working methods for both groups are clearly established and much discussion between formal face-to-face meetings is conducted through electronic means. Project MESA provides a kind of 'bicameral' model for a constructive technology assessment process that national governments might consider adopting as a strategy to promote closer integration among user groups and technical experts for the management of critical infrastructure, particularly within a mitigation-oriented policy framework, where a careful balance between the ambitious visions of application developers and the temperate judgements of technical experts is extremely important to ensure orderly growth and change over the long run. This balance is all the more important because incremental developments in large technical systems will often have serious path dependent consequences in the long run, hence the CWTA membership's concern with 'stranded investments' in the technical trials of Wireless E9-1-1. User groups may be unaware of such implications or unable to participate in proceedings that deal with such matters, owing to various constraints including limited staff expertise and financial resources. If communities of users are to participate more effectively in the assessment of technology projects, they must be made aware of important developments and moreover, they must be provided with access to the expertise necessary for them to comprehend how such developments may come to affect their interests in the future. Finally, these communities must in some way be enabled to discover and articulate their demands for new and/or improved applications and services that contribute to the objectives of mitigation-oriented policy initiatives.

Only if specific cases for relief have been identified will business groups or others likely be permitted to file interventions with the regulatory agency. At present, such cases may be difficult to recognize or establish simply because many user groups remain unaware of the possibilities for new services available to them. The government-sponsored forum described above would therefore in part be intended to raise awareness among user groups, to encourage demand articulation for new services, and to provide grounds for regulatory intervention when important network elements or other interconnections arrangements are not made available on fair terms and conditions to service providers.

Innovation with Targeted Support

The last point concerning the importance of ensuring interconnection on fair terms and conditions is notable because the case study also found that industry stakeholders might be willing to foster innovation in new technology and services but they were far less inclined to simply let the products of these efforts become widely available. This observation is supported by Mansell's 'strategic model' of innovation as described in chapter five, which suggests that dominant stakeholders will attempt to influence network design through delay or refusal of access to certain network elements. The CWTA, for instance, proved that it could encourage industry and other stakeholder cooperation in the conduct of successful technical trials for Wireless E9-1-1. Yet commercial deployment of Wireless E9-1-1 service proved much slower and a number of wireless carriers were reluctant to issue public commitments to offer the service even after it had been tested and became operational. Incumbent carriers operating outside the provinces in which technical trials took place were also reluctant to commit to upgrading the infrastructure or providing the network access services necessary to enable wireless CLECs such as Microcell to meet regulatory obligations with respect to E9-1-1 service in other provinces.

To counter such tendencies and to foster innovation and experimental designs, third party investors must have some assurance of timely deployment of the products of their efforts through fair terms and conditions of interconnection to the wider support infrastructure. The case study also suggests, however, that the culture of forbearance that has accompanied much regulatory reform may make regulatory agencies reluctant to become directly involved in the development and deployment of new services. My observation in the case of Wireless E9-1-1 is that the lack of direct supervision caused by the CRTC's failure to intervene in a more effective manner was partly responsible for many of the delays, added costs of proceedings, and ultimate uncertainty that may have had a significant impact not only on local competition but on the safety of consumers.

Innovation and experimentation in telecom services may be encouraged through appropriate incentive mechanisms provided by government programs directed at strategic niche management. In the wireless sector, for example, a portion of license fees from wireless service providers and future spectrum auctions could be made available to support research and development for new value-added services within a mitigation-oriented innovation program. The Canadian Wireless Telecommunications Association (CWTA) in its submission to the federal government's Innovation Strategy Consultation, has even argued for a trade-off in the form of reduced regulation costs 'as a means to promote reinvestment [and] encouraging more continuous innovation' (Canadian Wireless Telecommunications Association, 2002). Under a designated program of innovation, such reinvestment of revenues could be specifically targeted for research and development in value-added services for emergency management and business continuity planning. Qualifying projects could be promoted through or drawn from a government-sponsored reflexive forum much like the Project MESA working relationship described above, where a service specification group of users would provide input to a technical specification and design group of developers.

Enabling through Technical Expertise

Evidence from the case study also indicates that technology forcing is a difficult strategy with which to involve wide stakeholder participation, particularly when investigating a specific regulatory matter or when governments and regulatory agencies initiate public proceedings where the scope is limited to pre-defined issues based on past decisions. Although negotiated rule-making bodies such as the CRTC's Interconnection Steering Committee may be open to 'all interested parties,' they tend to operate in a consensus-driven context, often late in the design phase of a technology project, and they often focus on expediting highly technical matters rather than exploring a wide range of issues.

Yet, public policy objectives usually do establish a responsibility for regulatory agencies and other government departments to coordinate the community of users and the technical development of critical infrastructure to ensure adherence to competition directives and fulfilment of other public policy objectives (e.g., universal service). Likewise, regulatory agencies must now consider in what ways they are formally bound to mitigation-oriented policy initiatives through the social policy objectives contained in legislation or other policy directives. For the public information infrastructure in Canada, as I have noted, such a responsibility is established in federal legislation.

By way of taking action to fulfil this responsibility, the CRTC could, for instance, direct its Interconnection Steering Committee to form a working group to address and report on issues associated with disaster mitigation on a case-by-case basis, giving fair and timely consideration to requests made by new entrants and third party service providers. Such an arrangement could work in tandem with a program of strategic niche management to ensure that enabling technical requirements for the testing and deployment of new value-added services would be taken into consideration in a fair and timely manner. The specific focus of such a working group, however, should be to ensure that lower layers of interconnection space (physical transport and network access services) meet the minimum conditions needed to accommodate long-range prospects for mitigation-oriented activities. One task, for example, could be to establish baseline service capability requirements that would reasonably and universally extend within a country or region to ensure a close correspondence between mitigation initiatives and the orderly development of the infrastructure environment.

The Australian Communications Authority (ACA) has been involved in an initiative that shares features with jus such an enabling strategy. In 2001, the ACA convened an industry task force to prepare guidelines 'intended to materially assist in ensuring optimal communications support for emergency management situations' (Australian Communications Authority, 2001). The task force examined the problem and opportunity of providing a full range of services to emergency management organizations within a competitive telecom environment:

> Following the liberalisation of the telecommunications market in Australia and the advent of competition, the supply arrangements for telecommunications services has become more complex. Total reliance on Telstra [the incumbent carrier] for emergency communications management support is no longer appropriate. While Telstra remains the key provider for telecommunications services to ESOs this is not necessarily the case across the board. Other providers may appropriately be involved in the provision of pre-planned services where they are providing the basic telecommunications services to an ESO. Further, it is important that in the provision of ad hoc services, ESOs have potential access (via their pre-planned provider) to the full range of telecommunications that may be potentially available to meet an emergency situation—whether or not they are directly available from the pre-planned [primary] service provider. (Australian Communications Authority, 2001, p. 2)

The task force produced a set of protocols within a guideline document to describe the process by which ad hoc services were to be arranged between secondary and primary service providers (Australian Communications Industry Forum, 2002). The guideline document works within the competitive policy framework to maximize the opportunity for the emergency management community to benefit from innovation and advanced services deployed in the telecommunications infrastructure.

An example from a case in Canada that may have some relevance to the importance of such pre-established measures was filed with the CRTC in January 2002 as a Part VII application, not unlike the one involving Microcell and Wireless E9-1-1 (Canadian Radio-television and Telecommunications Commission, 2002). In this case, a wireline-based CLEC claimed that an incumbent carrier was providing discriminatory repair times for services over the local loops (owned by the incumbent and leased to the CLEC). This claim is principally about competitive disadvantage stemming from a quality of service issue, where the incumbent carrier appears to be lowering its standards for competitors who must also use those facilities. If we consider this case in the context of value-added services, such discriminatory practices could prove to be major barriers to the success of emerging mitigation-oriented innovations. In fact, it is conceivable that similar cases will continue to be filed in future with respect to value-added services provided by third parties, wireless carriers, CLECs, or even other incumbents, particularly where bottleneck facilities are used. By implementing guidelines similar to the Australian model, however, such cases might be circumvented at the outset.

Yet, ACA's initiative is a not a mitigation project in the strict sense adopted for this study because it does not address the process of growth and change itself.

It does invite a mitigation project insofar as operational effectiveness of any *ad hoc* services are constrained by the effective limitations of interconnection between primary and secondary service providers. This may be especially critical at the higher functional layers involving access to databases and other value-added services not otherwise regulated or standardized. In order to conform to 'mitigation' as it is understood through the Pressure and Release model, the ACA would need to initiate a task force for network interconnection arrangements as a pre-condition for developing, testing, and supplying ad hoc services between primary and secondary service providers during emergency situations.

Nonetheless, the ACA guidelines might provide a model for other countries wherein the regulatory agency could direct (or create if necessary) the appropriate technical working groups to consider the development of a similar set of protocols. In the context of a mitigation-oriented policy framework, however, the mandate of such an undertaking should be expanded to include the development of baseline service capability requirements through appropriate interconnection arrangements to ensure optimal conditions for providing ad hoc services in emergency situations. In Canada, for instance, such a task could be assigned to a specific working group within the CISC, which would in turn establish a formal decision-making process for the assessment of such enabling projects from a detailed technical standpoint.

Coordinated Intervention

Each of the proposed approaches described above can be summarized and plotted on the intervention matrix developed for this study. From the Wireless E9-1-1 case, I observed a counterproductive sequencing in the structure of interventions. In response to this observation, the proposed approaches provide a coordinated intervention strategy that begins with demand articulation then proceeds to problem formulation and on to design proposition. Communities of users are invited to articulate their needs and concerns in functional terms that correspond to the upper layers of interconnection space. Through dialogue in relatively non-technical language, this provides a shared learning experience between users and the technical experts who support demand articulation.. The results of the consultations are then translated into more specific problem formulations and design propositions as the structures of intervention shift to actuation through innovation and eventually to the lower layers, where technical details are worked out by specialist committees and working groups.

For instance, a reflexivity strategy based on wide stakeholder consultation through a government-sponsored forum for disaster mitigation should be the first point of contact intended to encourage discussion and such a forum would span all aspects of interconnection space, from lower physical layers through to value-added services and information content. Congruency among stakeholders at this early stage is founded on *learning* and can be supported through participation guidelines similar to those developed for Project MESA. The output of this stage could then feed into programs to support strategic niche management through targeted funding programs or other incentives. Many of the innovation activities supported through these programs would be value-added service concepts that reside in upper layers of interconnection space with requirements for lower layer interconnection arrangements. The output of these programs could drive sponsorship of tasks within technical working groups, where lower layer interconnection arrangements and long-range network planning would be addressed around the congruent notion of *enabling* innovation and experimentation in value-added services.

The intervention matrix presents a strategy to structure an evolving mix of stakeholders for recursive learning. It therefore can also be used to inform political questions raised in constructive technology assessment about stakeholder representation. A coordinated intervention strategy suggests that stakeholder representation may best proceed from a public forum within a government-supported locus of reflexivity, to a more specialized form of industry-led participation as the process moves to strategic niche management, then finally to specific task assignments within government/industry partnership under a specialized technical working group, like that of the CRTC's Interconnection Steering Committee in Canada or the technical specification group in Project MESA.

Table 7.7 Coordinated intervention strategies

	Reflexivity	Niche Management	Forcing
Legitimacy	Government-sponsored public forum	Membership in industry working group	Recognized qualifications to participate in technical working group
Congruency	Learning ⟹	Innovation ⟹	Enabling
Inscription	Participation Guidelines	Financial incentives designated for specific areas of innovation	Task assignments from regulatory agency or interested parties
Interconnection Space	All layers considered, with emphasis on user needs	Value-added services; Information content	Physical infrastructure; Network service

Under the modified structural arrangement as described above and summarized in Table 7.7, the initial learning stage becomes exploratory and relatively non-committal on the demand-side of the equation. As demand articulation develops into more targeted problem formulation and then into specific design propositions through strategic niche management, learning modulates between demand-side and supply-side perspectives. Finally, when design propositions are taken up within technical working groups, learning shifts to a predominantly supply-side discourse and provides policymakers and industry alike with early insight on path dependency and the viability of technology forcing strategies.

Looking Further Afield

I began this book with a call to extend our thinking about the concept of mitigation. Notwithstanding the volumes written on it previously, mitigation in my view has remained an ambitious but *ambiguous* idea that needs to be more clearly understood so it can stand apart from other activities in disaster management. In order to achieve this clarity, I have adopted the Pressure and Release Model and applied it to the study of critical infrastructure, suggesting that mitigation-oriented policy research be directed at the fundamental processes and influences on network development. This insight informs a central argument taken up in this book:

> Policymakers and researchers must understand growth and change in critical infrastructure if they are to effectively intervene in that process to better coordinate it with mitigation-oriented policy initiatives.

Based on this formative claim, the subsequent chapters of the book were directed toward understanding those fundamental forces involved in shaping the evolution of critical infrastructure—the 'root causes' and 'dynamic pressures' of growth and change in large technical systems. It has been my intention to use this shift in perspective to introduce a new way of thinking about the management of critical infrastructure along three lines of thinking: theoretical, methodological, and empirical.

Theoretically, the shift in perspective is toward a growing body of literature that recognizes risk and vulnerability as socially constructed conditions, intimately linked with wider forces of community development. The case study demonstrates the possibility of a fruitful theoretical synthesis and empirical application across scholarly domains within the reflexive tradition of science and technology studies, particularly within Bijker's theory of socio-technical change, Actor Network Theory, and historical studies in Large Technical Systems. This synthesis presents a theoretical account of technology dynamics that can be rendered through the three generic intervention strategies of Constructive Technology Assessment, resulting in a basic analytic framework for studying technology projects under development. Of course, more work needs to be done in the theoretical domain, particularly in understanding the design nexus as a rhetorical process involving analytic, political, and normative dimensions (as was suggested in chapter two).

Methodologically, the shift in perspective demonstrates the important contribution that the reflexive tradition of science and technology studies and technology assessment can bring to a more administratively oriented body of policy literature. The intervention matrix is a useful contribution to the study of infrastructure development, particularly in the application of 'interconnection space' as a third dimension of analysis, and it provides a useful tool for identifying and assessing opportunities for intervention in the complex and dynamic processes of a technology project in progress. Further development of the methodology with respect to the political and normative dimensions of CTA is needed, as is further refinement of the analytical undertaking developed for this book.

Much of the refinement of this methodology will come from additional empirical research. This book provides a preliminary sample, by applying the methodology to analyze a single instance of growth and change in public safety telecommunications. Results from the analysis offer a number of specific insights for structuring intervention to achieve public policy objectives. Further empirical studies are necessary, however, particularly comparative studies across critical infrastructure systems to enhance our understanding of growth and change in contemporary large technical systems. Better knowledge in turn can yield more appropriate intervention strategies

While this study has been limited to an analytic undertaking within the Constructive Technology Assessment approach, other undertakings are also possible. Further research on political and normative undertakings may be worth consideration or even necessary. For instance, this study did not examine specific political concerns associated with participation in planning forums nor did it attempt to identify all possible parties that may have a stake in the management of critical infrastructure. Further work could seek to identify key stakeholder groups in order to conduct research to better understand in detail the challenges of participation and interaction in various forums. A study on the normative aspects of critical infrastructure development could look into the philosophy or political economy of technology to reflect more deeply on the concept of risk as it pertains to the evolution of critical infrastructure and large technical systems more generally. Such work could critically reflect upon the fundamental assumptions that underpin current policy and regulation for critical infrastructure, especially with respect to public policy objectives established in legislation for telecommunications and other interdependent critical systems.

In the opening pages of the book I raised the spectre of the interdependency dilemma as a source of acute risk in advanced industrial societies. The appearance of mitigation-oriented policy in recent years clearly reflects a concern with interdependent systems, yet forward-thinking policy research remains constrained by inadequate conceptual approaches to the problem. This book has been written in hope of contributing something of significance not only to a re-orientation in thinking about the origin of the problem but, perhaps more importantly, to a dialogue about democratic solutions to achieve sustainable societies for future generations.

Bibliography

Abbate, J. (1994). The Internet Challenge: Conflict and Compromise in Computer Networking. In J. Summerton (Ed.), *Changing Large Technical Systems* (pp. 193-210). Boulder, CO: Westview Press.

Akrich, M. (1995). User Representations: Practices, Methods and Sociology. In A. Rip, T. Misa and *Schot* (Eds.), *Managing Technology in Society: The Approach of Constructive Technology Assessment* (pp. 167-184). New York: St. Martin's Press.

Alberta 9-1-1 Advisory Association. (1998, Dec. 16). Alberta Wireless E9-1-1 Trial Team: Minutes of Meeting #98-02. Available http://www.cwta.ca/

Alberta 9-1-1 Advisory Association. (1999a, Feb. 12). Alberta Wireless E9-1-1 Trial Team: Minutes of Meeting #99-03. Available http://www.cwta.ca/

Alberta 9-1-1 Advisory Association. (1999b, Mar. 11). Alberta Wireless E9-1-1 Trial Team: Minutes of Meeting #99-04. Available http://www.cwta.ca/

Alberta E9-1-1 Advisory Association. (2000a, May 22). Alberta Wireless 9-1-1 Trial Report: A Collaborative Effort to Enhance 9-1-1 Call Handling and Delivery from Wireless Phones. *Canadian Wireless Telecommunications Association (CWTA)*, 2001. Available http://www.cwta.ca/CWTASite/english/E911.html

Alberta E9-1-1 Advisory Association. (2000b, May 31). CRTC 8740-M29-0002/00 - Microcell Connexions Inc. - TN 2 - General Tariff - Rates and Conditions for Services E9-1-1 - 2000/05/31 - Public Safety Answering Point.

Alberta E9-1-1 Advisory Association. (2001, Jan. 3). CRTC 8740-T42-327/00 - TCI TN 327 and TN 327/A - General Tariff - Wireless Service Provider Enhanced Provincial 9-1-1 Network Access Service - 2001/01/03 - Alberta E9-1-1 Advisory Association. Available http://www.crtc.gc.ca//8740/eng/2000/t42-327.htm

Alberta E9-1-1 Advisory Association. (2002, Jan. 28). CRTC 8669-C12-01/01 - Public Notice 2001-110 - Conditions of service for wireless competitive local exchange carriers and for 9-1-1 services offered by wireless service providers - Reply Comments - Phase II - 2001/01/28 - Alberta E9-1-1 Advisory Association.

Aliant Telecom Inc., Bell Canada, MTS Communications Inc. and Saskatchewan Telecommunications. (2001a, Apr. 12). CRTC 8669-M16-01/01 - Microcell Telecommunications Inc. - Applications for Mandated Provision of Wireless Enhanced 9-1-1 Network Access Service - 2001/04/12 - Aliant Telecom Inc., Bell Canada, MTS Communications Inc., and Saskatchewan Telecommunications, (collectively the Companies).

Aliant Telecom Inc., Bell Canada, MTS Communications Inc. and Saskatchewan Telecommunications. (2001b, May 30). CRTC 8669-M16-01/01 - Microcell Telecommunications Inc. - Applications for Mandated Provision of Wireless Enhanced 9-1-1 Network Access Service - 2001/05/30 - Aliant Telecom Inc. (Aliant), Bell Canada, MTS Communications Inc. (MTS) and Saskatchewan Telecommunications (SaskTel).

Aliant Telecom Inc., Bell Canada, MTS Communications Inc. and Saskatchewan Telecommunications. (2001c, June 6). CRTC 8669-M16-01/01 - Microcell Telecommunications Inc. - Applications for Mandated Provision of Wireless Enhanced 9-1-1 Network Access Service - 2001/06/06 - Aliant Telecom Inc., Bell Canada, MTS Communications Inc. and Saskatchewan Telecommunications.

Anderson, P. S. and Gow, G. A. (2000). Commercial Mobile Telephone Services and the Canadian Emergency Management Community: Prospects and Challenges for the Coming Decade. *Office of Critical Infrastructure Protection and Emergency Preparedness (OCIPEP)*. Available http://www.ocipep-bpiepc.gc.ca

Anderson, P. S. and Stephenson, R. (1997). Disasters and the Information Technology Revolution. *Disasters: the Journal of Disaster Studies, Policy and Management*, 21(4).

Angus Telemanagement Group. (2001, Sept. 17). Telecom Update No. 300. Retrieved Oct. 2, 2001. Available http://www.angustel.ca/update/up300.html

Armstrong, J. E. and Harman, W. W. (1980). *Strategies for Conducting Technology Assessments*. Boulder: Westview Press.

Armstrong, M. (1998). Network Interconnection in Telecommunications. *The Economic Journal*, 108(May), 545-564.

Arnbak, J. C. (1997). Technology Trends and their Implications for Telecom Regulation. In Melody (Ed.), *Telecom Reform: Principles, Policies and Regulatory Practices* (pp. 67-82). Lyngby: Den Private Ingeniorfond, Technical University of Denmark.

Australian Communications Authority. (2001, Oct. 12). Communication Support for Emergency Management (Draft). Retrieved Jun. 18, 2002. Available http://www.aca.gov.au/licence/carrier/CSEM/guidelines.htm

Australian Communications Industry Forum. (2002, April). Industry Guideline: Communications Support for Emergency Response. Retrieved June, 2002. Available http://www.acif.org.au

Australian Government. (2003). Natural Disaster Mitigation Programme. *Department of Transport and Regional Services*. Retrieved May, 2004. Available http://www.dotars.gov.au/naturaldisasters/index.aspx

Bauer, J. M. (2003). Normative foundations of electronic communications policy in the European Union. In J. Jordana (Ed.), *Governing Telecommunications and the New Information Society in Europe* (pp. 110-133). Cheltenham: Edward Elgar.

BC 9-1-1 Service Providers Association. (2001a, Jan. 28). CRTC 8669-C12-01/01 - Public Notice 2001-110 - Conditions of service for wireless competitive local exchange carriers and for 9-1-1 services offered by wireless service providers - Reply Comments - Phase II. Available http://www.crtc.gc.ca/PartVII/Eng/2001/8669/C12-01.htm

BC 9-1-1 Service Providers Association. (2001b, Jan. 8). CRTC 8740-T42-327/00 - TCI TN 327 and TN 327/A - General Tariff - Wireless Service Provider Enhanced Provincial 9-1-1 Network Access Service - 2001/01/08 - British Columbia E-911 Service Providers Association - Revised Comments (010108_2.doc).

Bell Canada. (2000a, June 5). CRTC 40-M29-0002/00 - Microcell Connexions Inc. - TN 2 - General Tariff - Rates and Conditions for Services - 2000/06/05 - Bell Canada, Island Telecom Inc., Maritime Tel and Tel Limited, MTS Communications Inc., NBTel Inc. and NewTel Communications Inc.

Bell Canada. (2000b, July 11). CRTC 8740-C46-0001/00 - Clearnet PCS Inc. - TN 1 - 2000/07/11 - Bell Canada, Island Telecom Inc., Maritime Tel and Tel Limited, MTS Communications Inc., NBTel Inc. and NewTel Communications Inc. Available http://www.crtc.gc.ca/8740/eng/2000/c46-1.htm

Bell Canada. (2001a, Nov. 20). CRTC 8740-B2-6629/01 - Bell Canada - TN 6629 - Special Facilities Tariff - Wireless Service Provider Enhanced 9-1-1 Service (WSP E9-1-1). Available http://www.crtc.gc.ca/8740/eng/2001/b2-6629.htm

Bell Canada. (2001b, June 15). CRTC CISC - Intercarrier Operations Group - Emergency Services (9-1-1) - ESCO166.doc - Formely ESCOX166.doc - Clarification of 9-1-1 Street Address Guide (9-1-1 SAG) Provisioning to Wireless CLECs.

Bell Canada. (2002, Jan. 9). CRTC 8740-B2-6629/01 - Bell Canada - TN 6629 - Special Facilities Tariff - Wireless Service Provider Enhanced 9-1-1 Service (WSP E9-1-1) - 2002/01/09 - Bell Canada (Reply Comments).

Bell Mobility. (2001, Apr. 12). CRTC 8669-M16-01/01 - Microcell Telecommunications Inc. - Applications for Mandated Provision of Wireless Enhanced 9-1-1 Network Access Service - 2001/04/12 - Bell Mobility.

Benhabib, S. (1996). Toward a Deliberative Model of Democratic Legitimacy. In S. Benhabib (Ed.), *Democracy and Difference: Contesting the Boundaries of the Political* (pp. 67-94). Princeton: Princeton University Press.

Bereano, P. L. (1997). Reflections of a Participant-Observer: The Technocratic/Democratic Contradiction in the Practice of Technology Assessment. *Technological Forecasting and Social Change*, 54(2,3), 163-171.

Berloznik, R. and van Langenhove, L. (1998). Integration of Technology Assessment in R&D Management Practices. *Technological Forecasting and Social Change*, 58(1,2), 23-33.

Bernard, H. R. (2000). *Social Research Methods: Qualitative and Quantitative Approaches.* Thousand Oaks, CA: Sage.

Bijker, W. (1995). *Of Bicycles, Bakelite, and Bulbs: Toward a Theory of Sociotechnical Change.* Cambridge: MIT Press.

Bijker, W. and Law, J. (Eds.). (1994). *Shaping Technology/Building Society: Studies in Sociotechnical Change.* Cambridge, MA: MIT Press.

Blackwell, R., Craig, S. and Bell, A. (1999, July 17). The Phone Crisis: Bank machines, credit cards knocked out. *The Globe and Mail*, p. A6.

Blaikie, P., Cannon, T., Davis, I. and Wisner, B. (1994). *At Risk: Natural Hazards, People's Vulnerability, and Disasters.* London: Routledge.

Bruce, J. (1999). Disaster loss mitigation and sustainable development. In J. Ingleton (Ed.), *Natural Disaster Management: A Presentation to Commemorate the International Decade for Natural Disaster Reduction (IDNDR)* (pp. 28-30). Leicester: Tudor Rose.

Buchanan, R. (1987). Declaration by Design: Rhetoric, Argument, and Demonstration in Design Practice. In V. Margolin (Ed.), *Design Discourse: History, Criticism, Theory* (pp. 91-109). Chicago: University of Chicago Press.

Buchanan, R. (1996). Wicked Problems in Design Thinking. In V. Margolin and R. Buchanan (Eds.), *The Idea of Design: A Design Issues Reader* (pp. 3-20). Cambridge: MIT Press.

Callon, M. and Latour, B. (1981). Unscrewing the Big Leviathan: How Actors Macro-Structure Reality and How Sociologists Help Them to Do So. In K. Knorr-Cetina and A. V. Cicourel (Eds.), *Advances in Social Theory and Methodology* (pp. 277-303). London: Routledge and Kegan Paul.

Callon, M. and Law, J. (1997). After the Individual in Society: Lessons on Collectivity from Science, Technology and Society. *Canadian Journal of Sociology*, 22(2), 165-182.

Canada Department of Justice. (2001, Aug. 31). Telecommunications Act (1993, c.38).

Canada. (2000, June). Senate Standing Committee on National Finance. Report on the Committee's Examination of Canada's Emergency and Disaster Preparedness.

Canada. (2001, Feb. 5). Office of Critical infrastructure Protection and Emergency Preparedness. Department of National Defence. Retrieved July, 2001. Available http://www.dnd.ca/eng/archive/2001/feb01/06protect_b_e.htm

Canada. (2004). *National Emergencies: Canada's Fragile Front Lines.* Ottawa: Senate Standing Committee on National Security and Defence.

Canadian Radio-television and Telecommunications Commission. (1996a). Telecom Decision 96-14: Regulation of Mobile Wireless Telecommunications Services.

Canadian Radio-television and Telecommunications Commission. (1996b, Aug. 1). Telecom Public Notice CRTC 96-28: Implementation of Regulatory Framework - Development of Carrier Interfaces and Other Procedures.

Canadian Radio-television and Telecommunications Commission. (1997). Telecom Decision 97-8: Local Competition.

Canadian Radio-television and Telecommunications Commission. (1999, Dec. 8). Telecom Order 99-1127: Microcell's tariff application to provide local services.

Canadian Radio-television and Telecommunications Commission. (2000a, Sept. 8). Order CRTC 2000-830 - General Tariff approved on an interim basis with modifications for Clearnet PCS Inc.

Canadian Radio-television and Telecommunications Commission. (2000b, Sept. 8). Order CRTC 2000-831: General Tariff approved on an interim basis with modifications for Microcell Connexions Inc.

Canadian Radio-television and Telecommunications Commission. (2001a). CRTC 8669-M16-01/01 - Microcell Telecommunications Inc. - Applications for Mandated Provision of Wireless Enhanced 9-1-1 Network Access Service - 2001/05/16 - Commission Letter (1 of 2).

Canadian Radio-television and Telecommunications Commission. (2001b). CRTC 8669-M16-01/01 - Microcell Telecommunications Inc. - Applications for Mandated Provision of Wireless Enhanced 9-1-1 Network Access Service - 2001/05/16 - Commission Letter (2 of 2).

Canadian Radio-television and Telecommunications Commission. (2001c, Feb. 16). CRTC 8698-C12-13/01 - Enhanced 9-1-1 Wireless Trial - Province of Ontario - 2001/02/16 - Commission Letter.

Canadian Radio-television and Telecommunications Commission. (2001d, Mar. 31). CRTC Interconnection Steering Committee Administrative Guidelines (Version 1.1).

Canadian Radio-television and Telecommunications Commission. (2001e, Oct. 12). CRTC Telecommunications Rules of Procedure.

Canadian Radio-television and Telecommunications Commission. (2001f, Aug. 9). Decision CRTC 2001-475: Allocation of three-digit dialing for public information and referral services.

Canadian Radio-television and Telecommunications Commission. (2001g, Feb. 2). Order CRTC 2001-97 - Wireless service provider enhanced provincial 9-1-1 network access service.

Canadian Radio-television and Telecommunications Commission. (2001h, Dec. 21). Order CRTC 2001-902 - Wireless Service Provider Enhanced 9-1-1 Service.

Canadian Radio-television and Telecommunications Commission. (2002, Jan. 23). CRTC 8622-C25-14/02 - Call-Net Enterprises - Requesting an order to require ILECs to file a tariff for 4-Hour Mean Time to repair on unbundled local loops.

Canadian Radio-television and Telecommunications Commission. (2003, Aug. 12). Decision CRTC 2003-53: Conditions of service for wireless competitive local exchange carriers and for emergency services offered by wireless service providers.

Canadian Wireless Telecommunications Association. (2002, October). Innovation in Canada's Wireless Industry: Maintaining the Momentum.

Canadian Wireless Telecommunications Association. (2003a, Sept.). E9-1-1. *CWTA: Health and Safety.* Available http://www.cwta.ca/CWTASite/english/E911.html

Canadian Wireless Telecommunications Association. (2003b). Mobile Wireless Subscribers in Canada. *CWTA Facts.* Available http://www.cwta.ca/industry_guide/facts.php3

Canadian Wireless Telecommunications Association, Wireless E9-1-1 Working Group. (1997a). June 17th Round-Table Discussion: Wireless 9-1-1 Service Summary of Discussion. Available http://www.cwta.ca/CWTASite/english/E911.html

Canadian Wireless Telecommunications Association, Wireless E9-1-1 Working Group. (1997b). Minutes of July 29, 1997 Meeting. Retrieved Nov. 7, 2001. Available http://www.cwta.ca/CWTASite/english/E911.html

Canadian Wireless Telecommunications Association, Wireless E9-1-1 Working Group. (1997c). Minutes of Meeting of November 10, 1997. Retrieved Nov. 7, 2001. Available http://www.cwta.ca/CWTASite/english/E911.html

Canadian Wireless Telecommunications Association, Wireless E9-1-1 Working Group. (1998a). Minutes of Meeting of April 24, 1998. Retrieved Nov. 7, 2001. Available http://www.cwta.ca/CWTASite/english/E911.html

Canadian Wireless Telecommunications Association, Wireless E9-1-1 Working Group. (1998b). Minutes of Meeting of January 23, 1998. Retrieved Nov. 7, 2001. Available http://www.cwta.ca/CWTASite/english/E911.html

Canadian Wireless Telecommunications Association, Wireless E9-1-1 Working Group. (1998c). Minutes of Meeting of March 20, 1998. Retrieved Nov. 7, 2001. Available http://www.cwta.ca/CWTASite/english/E911.html

Canadian Wireless Telecommunications Association, Wireless E9-1-1 Working Group. (1998d). Minutes of Meeting of September 28, 1998. Retrieved Nov. 7, 2001. Available http://www.cwta.ca/CWTASite/english/E911.html

Canadian Wireless Telecommunications Association, Wireless E9-1-1 Working Group. (2000). Minutes of Meeting of November 14, 2000. Retrieved Nov. 7, 2001. Available http://www.cwta.ca/CWTASite/english/E911.html

Canadian Wireless Telecommunications Association, Wireless E9-1-1 Working Group. (2001a). Meeting of Minutes of July 11, 2001. Retrieved Nov. 7, 2001. Available http://www.cwta.ca/CWTASite/english/E911.html

Canadian Wireless Telecommunications Association, Wireless E9-1-1 Working Group. (2001b). Meeting of Minutes of October 11, 2001. Retrieved Apr. 8, 2002. Available http://www.cwta.ca/CWTASite/english/E911.html

Canadian Wireless Telecommunications Association, Wireless E9-1-1 Working Group. (2001c). Minutes of Meeting of February 15, 2001. Retrieved Nov. 7, 2001. Available http://www.cwta.ca/CWTASite/english/E911.html

Canadian Wireless Telecommunications Association, Wireless E9-1-1 Working Group. (2001d). Minutes of Meeting of May 8, 2001. Retrieved Nov. 7, 2001. Available http://www.cwta.ca/CWTASite/english/E911.html

Cattaneo, C. (2000, Nov. 29). U.S. cable mishap disrupts CDNX. *National Post Online.* Retrieved Dec. 5, 2000. Available http://www.nationalpost.com

Cheney, P. (1999, July 17). Phones go dead, Toronto put on hold. *The Globe and Mail,* p. A1+.

City of Kobe. (2000, Apr. 19). Post-Quake Citizen Support Services Head Office. Record of Kobe's Post-Quake Socioeconomic Rehabilitation: Five Years after the Great Hanshin-Awaji Earthquake.

Clearnet PCS Inc. (2000, June 2). CRTC 8740-C46-0001/00 - Clearnet PCS Inc. - TN 1 - General Tariff.

Communauté urbaine de Montréal and Union des Municipalités du Québec. (2001, Apr. 12). CRTC 8669-M16-01/01 - Microcell Telecommunications Inc. - Applications for Mandated Provision of Wireless Enhanced 9-1-1 Network Access Service.

Cordesman, A. H. (2002). Cyber-threats, Information Warfare, and Critical Infrastructure Protection: Defending the U.S. Homeland. Westport, Conn: Praeger.

Crondstedt, M. (2002). Prevention, Preparedness, Response, Recovery—an outdated concept? *Australian Journal of Emergency Management,* 17(2), 10-13.

CRTC Interconnection Steering Committee. (1999, Nov. 12). Network Security Working Group - NSTF0006 - PSAP emergency ALI database lookup for non 9-1-1 dialled emergency calls or for 9-1-1 calls where there is no ALI record and SPID is required.

CRTC Interconnection Steering Committee. (2000, Nov. 27). CRTC CISC - Intercarrier Operations Group - Emergency Services (9-1-1) Contributions - ESCO156.doc - Formely ESCOX156.doc - Wireless CLEC Arrangements and 9-1-1 Trunk-Side CLEC Interconnection Document (Release 3.4).

CRTC Interconnection Steering Committee. (2001, Aug. 17). CRTC CISC - Intercarrier Operations Group - Emergency Services (9-1-1) ESTF029.doc - Wireless CLEC - Amendments to ESWG Trunk-Side CLEC Interconnection Document (TIF 29).

Cutcliffe, S. (2000). Ideas, Machines, and Values: An Introduction to Science, Technology, and Society Studies. Lanham: Rowman and Littlefield.

Darlington, J. and Simpson, D. (2001). Envisioning Sustainable Communities: The Question of Disasters. *Natural Hazards Review*, 2(2), 43-44.

David, P. (1986). Understanding the Economics of Qwerty: the Necessity of History. In W. N. Parker (Ed.), *Economic History and the Modern Economist* (pp. 30-49). Oxford: Blackwell.

Davies, A. (1994). *Telecommunications and Politics: The Decentralised Alternative*. London: Pinter.

Economides, N. (1996). The Economics of Networks. *International Journal of Industrial Organization*, 14(2), n/a.

Edge, D. (1995). The Social Shaping of Technology. In N. Heap, R. Thomas, G. Einon, R. Mason and H. McKay (Eds.), *Information Technology and Society: A Reader* (pp. 14-32). London: SAGE.

Eijndhoven, J. v. (1997). Technology Assessment: Process or Product? *Technological Forecasting and Social Change*, 54(2,3), 269-286.

Emergency Preparedness Canada. (1995). *Departmental Planning Responsibilities for Emergency Preparedness*. Ottawa: Government of Canada.

Emergency Preparedness Canada. (1999). The Story of Emergency Preparedness Canada.

European Telecommunications Standards Institute. (2003). EMTEL—Emergency Telecommunications. Available http://www.emtel.etsi.org/overview.htm

Federal Emergency Management Agency. (1996). Basic Principles of the National Mitigation Strategy. *Mitigation: Reducing Risk Through Mitigation*. Retrieved April, 2001. Available http://www.fema.gov/mit/prncpl.htm

Feenberg, A. (1999). *Questioning Technology*. London: Routledge.

Frieden, R. (2002, Sept. 29). Adjusting the Horizontal and Vertical in Telecommunications Regulation: A Comparison of the Traditional and a New Layered Approach. Paper presented at the Telecommunications Policy Research Conference, Alexandria, Virginia.

Geis, D. (2000). By Design: The Disaster Resistant and Quality-of-Life Community. *Natural Hazards Review*, 1(3), 151-160.

Gheorghe, A. V. (2004). Risks, vulnerability, sustainability and governance: a new landscape for critical infrastructures. *International Journal of Critical Infrastructures*, 1(1), 118-124.

Gilbert, C. (1998). Studying Disaster: Changes in the main conceptual tools. In E. Quarentelli (Ed.), *What is a Disaster? Perspectives on the Question* (pp. 11-18). New York: Routledge.

Grieve, W. (2000, Apr. 18). Constitutional Structure and Regulation of Telecommunications in Canada. The Law Review of Michigan State University-Detroit College of Law. Retrieved Oct. 19, 2001. Available http://www.dcl.edu/lawrev/2000-1/Grieve.htm

Grin, J. and Graaf, H. v. d. (1996). Technology Assessment as Learning. *Science, Technology, and Human Values*, 21(1), 72-99.

Grunwald, A. (2000). Technology Policy between Long-Term Planning Requirements and Short-Ranged Acceptance Problems: New Challenges for Technology Assessment. In J. Grin and A. Grunwald (Eds.), *Vision Assessment: Shaping Technology in 21st Century Society: Towards a Repertoire for Technology Assessment* (pp. 99-148). New York: Spring Verlag.

Harbi, M. (2001, Nov. 16). *Disaster Mitigation via Telecommunications: The Tampere Convention*. Paper presented at the ITU Telecom Africa 2001 Policy and Development Summit, Johannesburg, South Africa.

Hawkins, R. (1995). The User Role in the Development of Technical Standards for Telecommunications. *Technology Analysis & Strategic Management*, 7(1), 21-40.

Hawkins, R. (1997). The Changing Nature of Technical Regulation in Telecom Networks. In Melody (Ed.), *Telecom Reform: Principles, Policies and Regulatory Practices* (pp. 197-206). Lyngby: Den Private Ingeniorfond, Technical University of Denmark.

Hedrick, T. E., Bickman, L. and Rog, D. J. (1993). *Applied Research Design: A Practical Guide*. London: SAGE.

Hellström, T. (2003). Systemic innovation and risk: technology assessment and the challenge of responsible innovation. *Telecommunications Policy*, 25, 369-384.

Herdman, R. C. and Jensen, J. E. (1997). The OTA Story: The Agency Perspective. *Technological Forecasting and Social Change*, 54(2,3), 131-143.

Hill, C. T. (1997). The Congressional Office of Technology Assessment: A Retrospective and Prospects for the Post-OTA World. *Technological Forecasting and Social Change*, 54(2,3), 191-198.

Hoffman, K. W. (1990). Report on the Survivability of the Canadian Telecommunications Infrastructure. Ottawa: Industry Canada.

Hughes, T. (1983). *Networks of power: electrification in Western society, 1880-1930*. Baltimore: Johns Hopkins University Press.

Hughes, T. P. (1987). The Evolution of Large Technological Systems. In W. Bijker, Hughes and T. Pinch (Eds.), *The Social Construction of Technological Systems: New Directions in the Sociology and History of Technology* (pp. 51-82). Cambridge: MIT Press.

Industry Canada. (2002, July). Emergency Telecommunications: Legal Responsibilities. Retrieved May, 2004. Available http://spectrum.ic.gc.ca/urgent/english/legal.html

Innis, H. A. (1951). *The Bias of Communication* (1991 ed.). Toronto: University of Toronto Press.

Innis, H. A. (1986). *Empire and Communications*. Victoria, BC: Press Porcepic.

Institute for Catastrophic Loss Reduction. (1998). A National Mitigation Policy. *Emergency Preparedness Canada*. Retrieved April, 2001. Available http://www.epc-pcc.gc.ca/hottopics/what_hot/old_mitiga.html

Insurance Bureau of Canada. (1999). A National Mitigation Strategy: Protecting Canadians from severe weather and earthquakes. *Ice Storm '98 Project*. Retrieved March, 2001. Available http://qsilver.queensu.ca/~icestudy/

Jeggle, T. (1999). The goals and aims of the Decade. In J. Ingleton (Ed.), *Natural Disaster Management: A Presentation to Commemorate the International Decade for Natural Disaster Reduction* (IDNDR) (pp. 24-27). Leicester: Tudor Rose.

Joerges, B. (1988). Large Technical Systems: Concepts and Issues. In R. Mayntz and Hughes (Eds.), *The Development of Large Technical Systems* (pp. 9-36). Boulder, CO: Westview Press.

Karlsson, M. and Sturesson, L. (Eds.). (1995). *The World's largest machine : global telecommunications and the human condition*. Stockholm: Almqvist and Wiksell International.

Kelly, C. (1999). Simplifying disasters: developing a model for complex non-linear events. *Australian Journal of Emergency Management*, 14(1), 25-27.

Kemp, R. J. (1997, Oct. 8). Conclusions SNM workshop in Utrecht. *European Commission - Joint Research Centre - Institute for Prospective Technological Studies*. Retrieved July, 2004. Available www.jrc.es/projects/ snm /utrecht3.rtf

Klein, H. K. and Kleinman, D. L. (2002). The Social Construction of Technology: Structural Considerations. *Science, Technology, and Human Values*, 27(1), 28-52.

La Porte, T. (1997). New Opportunities for Technology Assessment in the Post-OTA World. *Technological Forecasting and Social Change*, 54(2,3), 199-214.

Latour, B. (1993). *We Have Never Been Modern* (C. Porter, Trans.). Cambridge, MA: Harvard University Press.

Latour, B. (1995). Mixing Humans and Nonhumans Together: The Sociology of a Door-Closer. In S. L. Star (Ed.), *Ecologies of Knowledge: Work and Practices in Science and Technology* (pp. 257-277). Albany: SUNY Press.

Latour, B. (1996). *Aramis or the Love of Technology* (C. Porter, Trans.). London: Harvard University Press.

Latour, B. (1997). On Actor-Network Theory: A Few Clarifications. Centre for Social Theory and Technology, Keele University, *UK*. Retrieved January, 1998. Available http://www.keele.ac.uk/depts/stt/stt/ant/latour.htm

Latour, B. (1999). On Recalling ANT. In J. Law and J. Hassard (Eds.), *Actor Network Theory and After* (pp. 15-25). Oxford: Blackwell.

Law, J. (1999). After ANT: Complexity, Naming and Topology. In J. Law and J. Hassard (Eds.), *Actor Network Theory and After* (pp. 1-13). Oxford: Blackwell.

Lawson, P. (1998, Nov. 13). Procedural Fairness and Deregulation: A Consumer Advocate's View of the Experience with Telecommunications. *Public Interest Advocacy Centre (PIAC): Telecommunications*. Available http://www.piac.ca/cbasp.htm

Lessig, L. (1999). *Code and Other Laws of Cyberspace*. New York: Basic Books.

Leydesdorff, L. (1996). The Evolution of Communication Systems. *International Journal of Systems Research and Information Science*, 6, 219-230.

Leyten, J. and Smits, R. (1996). The role of technology assessment in technology policy. *International Journal of Technology Management*, 11(5/6), 688-702.

Loka Institute. (2004, Feb. 24). Loka's Vision. Available http://www.loka.org/

Lucent Technologies. (1999). Understanding Wireless E9-1-1. *APCO International's Wireless E9-1-1*. Available http://www.apco911.org/gov/wireless.html

Luhmann, N. (1995). *Social Systems* (J. B. Jr., Trans.). Stanford, CA: Stanford University Press.

Mambrey, P. and Tepper, A. (2000). Technology Assessment as Metaphor Assessment: Visions Guiding the Development of Information and Communications. In J. Grin and A. Grunwald (Eds.), *Vision Assessment: Shaping Technology in 21st Century Society: Towards a Repertoire for Technology Assessment* (pp. 33-52). New York: Spring Verlag.

Mansell, R. (1990). Rethinking the telecommunication infrastructure: The new 'black box'. *Research Policy*, 19(6), 501-515.

Mansell, R. (1993). *The New Telecommunications: A Political Economy of Network Evolution*. London: SAGE.

Mansell, R. (1996). Communication by Design? In R. Mansell and R. Silverstone (Eds.), *Communication by Design* (pp. 15-43). New York: Oxford University Press.

Mansell, R. (1999). Designing Networks to Capture Customers: Policy and Regulation Issues for the New Telecom Environment. In Melody (Ed.), *Telecom Reform: Principles, Policies and Regulatory Practices* (pp. 83-96). Lyngby: Den Private Ingeniorfond, Technical University of Denmark.

Mansell, R. and Silverstone, R. (Eds.). (1996). *Communication by Design*. New York: Oxford University Press.

Masera, M. and Wilikens, M. (2001, June 27-29, 2001). *Interdependencies with the information infrastructure: dependability and complexity issues*. Paper presented at the Fifth International Conference on Technology, Policy and Innovation, The Hague, Netherlands. Retrieved, from http://www.delft2001.tudelft.nl/

Mayntz, R. and Hughes, T. (Eds.). (1988). *The Development of Large Technical Systems*. Boulder, CO: Westview Press.

Melody, W. (1997). Interconnection: Cornerstone of Competition. In Melody (Ed.), *Telecom Reform: Principles, Policies and Regulatory Practices* (pp. 53-66). Lyngby: Den Private Ingeniorfond, Technical University of Denmark.

Melody, W. H. (1999). Telecom reform: progress and prospects. *Telecommunications Policy*, 23, 7-34.

Microcell Connexions Inc. (1998, May 28). Microcell: First Wireless Provider Moving to Implement Co-Carrier Status (Press Release).

Microcell Connexions Inc. (1999, May 26). CRTC 8740-M29-1/99 - Microcell - TN 1 and 1/A - General Tariff - Competitive Local Exchange Carrier (CLEC).

Microcell Connexions Inc. (2000a, May 3). CRTC 40-M29-0002/00 - Microcell Connexions Inc. - TN 2 - General Tariff - Rates and Conditions for Services.

Microcell Connexions Inc. (2000b, June 12). CRTC 40-M29-0002/00 - Microcell Connexions Inc. - TN 2 - General Tariff - Rates and Conditions for Services - 2000/06/12 - Microcell Telecommunications Inc. (Reply Comments).

Microcell Telecommunications Inc. (2001a, Dec. 14). CRTC 8669-C12-01/01 - Public Notice 2001-110 - Conditions of service for wireless competitive local exchange carriers and for 9-1-1 services offered by wireless service providers - Comments - 2001/12/14 - Microcell Telecommunications Inc.

Microcell Telecommunications Inc. (2001b, Mar. 13). CRTC 8669-M16-01/01 - Microcell Telecommunications Inc. - Applications for Mandated Provision of Wireless Enhanced 9-1-1 Network Access Service - 2001/03/13 - Microcell Telecommunications Inc. (Application).

Microcell Telecommunications Inc. (2001c, Apr. 23). CRTC 8669-M16-01/01 - Microcell Telecommunications Inc. - Applications for Mandated Provision of Wireless Enhanced 9-1-1 Network Access Service - 2001/04/23 - Microcell Telecommunications Inc. (Reply Comments).

Microcell Telecommunications Inc. (2001d, June 7). CRTC 8669-M16-01/01 - Microcell Telecommunications Inc. - Applications for Mandated Provision of Wireless Enhanced 9-1-1 Network Access Service - 2001/06/07 - Microcell Telecommunications Inc. (Reply Comments).

Microcell Telecommunications Inc. (2001e, Dec. 20). CRTC 8740-B2-6629/01 - Bell Canada - TN 6629 - Special Facilities Tariff - Wireless Service Provider Enhanced 9-1-1 Service (WSP E9-1-1) - 2001/12/20 - Microcell Telecommunications Inc.

Microcell Telecommunications Inc. (2001f, Feb. 1). CRTC CISC - Intercarrier Operations Group - Emergency Services (9-1-1) - ESCO159.doc - Formely ESCOX159.doc - Entry of Wireless End-User Subscriber Data into 9-1-1 Automatic Location Information (ALI) Databases.

Microcell Telecommunications Inc. (2001g, Mar. 13). Microcell to Deploy Wireless E9-1-1 in Alberta and British Columbia - Wireless company also asks CRTC to mandate E9-1-1 service in rest of Canada (Press Release).

Mindel, J. and Sirbu, M. (2000). Regulatory Treatment of IP Transport and Services. *MIT Program on Internet and Telecoms Convergence*. Retrieved July 23, 2002. Available http://itc.mit.edu/itel/docs/2000/mindel_sirbu_TPRC00.pdf

Misa, T., Brey, P. and Feenberg, A. (Eds.). (2003). *Modernity and Technology.* Cambridge, Mass: The MIT Press.

National Communications System. (2004a). Government Emergency Telecommunications Service (GETS). Retrieved May, 2004. Available http://gets.ncs.gov/

National Communications System. (2004b, April). National Communications System (homepage). *Department of Homeland Security.* Retrieved May, 2004. Available http://www.ncs.gov/index.html

National Emergency Number Association. (2002a). 9-1-1 Facts. Retrieved Mar. 30, 2002. Available http://www.nena9-1-1.org/PR_Publications/Devel_of_911.htm

National Emergency Number Association. (2002b, Jan. 31). 9-1-1 Tutorial. Retrieved Apr. 8, 2002. Available http://www.nena.org

National Emergency Number Association. (2003). Wireless E9-1-1 Overview. Retrieved June, 2004. Available http://www.nena.org/Wireless911/Overview.htm

National Emergency Number Association. (2004, Feb. 6). NENA/DOT Wireless Deployment Reports. Retrieved June, 2004. Available http://nena.ddti.net/Reports/

National Institute of Standards and Technology. (1995). *The Impact of FCC's Open Network Architecture on NS/EP Telecommunications Security* (Special Publications No. 800-11). Washington: US Department of Commerce.

National Security Telecommunications Advisory Committee. (2003, December). NSTAC (homepage). Retrieved May, 2004. Available http://www.ncs.gov/nstac/nstac.htm

Newton, J. (1997, Jan.). Federal Legislation for Disaster Mitigation - A Comparative Assessment Between Canada and the United States. Retrieved May, 2002. Available http://www.epc-pcc.gc.ca/research/scie_tech/en_mitigat/index_e.html

Newton, J. (2001, June 26). *Loss Reduction Through Mitigation: The Focus of the Future.* Paper presented at the 11th World Conference on Disaster Management, Hamilton, Ontario.

Nguyen, N. T., Lobet-Maris, C., Berleur, J. and Kusters, B. (1996). Methodological issues in information technology assessment. *International Journal of Technology Management,* 11(5/6), 566-580.

Nicolet, R. (1999, Apr. 7). Report by the Nicolet Commission. Ministère de la Securité publique du Québec. Available http://www.msp.gouv.qc.ca

Nigg, J. and Tierney, K. (1995). *Business Vulnerability and Disaster-Related Lifeline Disruption.* Newark, Delaware: Disaster Research Center, University of Delaware.

Noam, E. M. (2001). *Interconnecting the Network of Networks.* Cambridge, MA: The MIT Press.

Office of Critical Infrastructure Protection and Emergency Preparedness. (2001, Jun. 26). Government of Canada launches consultations on the development of a National Disaster Mitigation Strategy. Available http://www.ocipep-bciepc.gc.ca

Office of Critical Infrastructure Protection and Emergency Preparedness. (2002, Jan.). Towards a National Disaster Mitigation Strategy: Discussion Paper. Retrieved Feb., 2002. Available http://www.epc-pcc.gc.ca/publicinfo/NDMS/Files/Disc_e.pdf

Ontario 9-1-1 Advisory Board and Alberta 9-1-1 Advisory Association. (2001a, Jan. 8). CRTC CISC - Intercarrier Operations Group - Emergency Services (9-1-1) - ESCO158.doc - Formely ESCOX158.doc - 9-1-1 Wireless CLEC Subscriber Data Requirements - Ontario 9-1-1 and AEAA.

Ontario 9-1-1 Advisory Board and Alberta 9-1-1 Advisory Association. (2001b, Mar. 21). CRTC CISC - Intercarrier Operations Group - Emergency Services (9-1-1) - ESCO162.doc - Formely ESCOX162.doc – 9-1-1 Wireless CLEC Subscriber Data Requirements - Ontario 9-1-1 and AEAA.

Ontario 9-1-1 Advisory Board and Communaute urbaine de Montreal. (2001, June 6). CRTC CISC - Intercarrier Operations Group - Emergency Services (9-1-1) - ESCO165.doc - Formely ESCOX165.doc - SAG Availability Wireless CLECs - Bell Canada Serving Territory - Ontario 9-1-1 and CUM/UMQ.

Ontario 9-1-1 Advisory Board. (1997, May 20). Letter to CRTC Addressing Decision 97-8 and Decision 97-9.

Ontario 9-1-1 Advisory Board. (2001, Feb. 28). CRTC CISC - Intercarrier Operations Group - Emergency Services (9-1-1) - ESCO160.doc - Formely ESCOX160.doc - Wireless CLEC: Amendments to ESWG Trunk-Side CLEC Interconnection Document - Ontario 9-1-1. Retrieved Jan. 15, 2001. Available http://www.crtc.gc.ca/cisc/eng/cisf3e4b.htm

Ontario E9-1-1 Wireless Trial Committee. (2000, Oct. 20). Minutes of Meeting, October 20. *Canadian Wireless Telecommunications Association*. Available http://www.cwta.ca

Organisation for Economic Co-operation and Development. (2002). OECD Guidelines for the Security of Information Systems and Networks: Towards a Culture of Security. Directorate for Science, Technology, and Industry. Available http://www.oecd.org/

Orlikowski, W. and Gash, D. (1994). Technological Frames: Making Sense of Information Technology in Organizations. *ACM Transactions on Information Systems*, 12(2), 174-207.

Pinch, T. and Bijker, W. (1987). The Social Construction of Facts and Artifacts: Or How the Sociology of Science and the Sociology of Technology Might Benefit Each Other. In W. Bijker, Hughes and T. Pinch (Eds.), *The Social Construction of Technological Systems: New Directions in the Sociology and History of Technology* (pp. 17-50). Cambridge: MIT Press.

Porter, A. L. (1980). *A Guidebook for Technology Assessment and Impact Analysis*. New York: North-Holland.

Project MESA. (2001). Project MESA (White Paper). Retrieved January, 2003. Available http://www.projectmesa.org/whitepaper/WhitePaper_MESA_0110.pdf

Project MESA. (2002, November). Project MESA Statement of Requirements. Retrieved January, 2003. Available http://www.projectmesa.org/SoR.htm

Project MESA. (2003, February). Project MESA: Mobile Broadband for Public Safety. Available http://www.projectmesa.org

Radder, H. (1996). Normative Reflexions on Constructivist Approaches to Science and Technology. In *In and About the World: Philosophical Studies of Science and Technology* (pp. 93-117). New York: State University of New York Press.

Ray, T. (2001, Sept. 21). Our Vulnerable Telecom System. *Yahoo! Finance*. Retrieved Oct. 2, 2001. Available http://biz.yahoo.com

Rip, A. (1994). Science and Technology Studies and Constructive Technology Assessment. *European Society for the Study of Science and Technology*, 13(3), not paginated.

Rip, A., Misa, T. and Schot, J. (Eds.). (1995). *Managing Technology in Society: The Approach of Constructive Technology Assessment*. New York: St. Martin's Press.

Robinson, C. P., Woodard, J. B. and Varnado, S. G. (1998, Fall). Critical Infrastructure: Interlinked and Vulnerable. *Issues in Science and Technology (Online)*. Retrieved May, 2004. Available http://www.issues.org/issues/15.1/robins.htm

Robinson, J. B., Francis, G., Lerner, S. and Legge, R. (1996). Defining a Sustainable Society. In J. B. Robinson (Ed.), *Life in 2030: Exploring a Sustainable Future for Canada* (pp. 26-52). Vancouver: UBC Press.

Rogers Wireless Inc. (2001, Dec. 19). CRTC 8740-B2-6629/01 - Bell Canada - TN 6629 - Special Facilities Tariff - Wireless Service Provider Enhanced 9-1-1 Service (WSP E9-1-1) -2001/12/19 - Rogers Wireless Inc.

Rowe, G. and Frewer, L. (2000). Public Participation Methods: A Framework for Evaluation. *Science, Technology and Human Values*, 25(1), 3-29.

Samarajiva, R. (2001, Nov. 16). *Disaster Preparedness and Recovery: A Priority fo Telecom Regulatory Agencies in Liberalized Environments.* Paper presented at the ITU Telecom Africa 2001 Policy and Development Summit, Johannesburg, South Africa.

Schiff, A. J. and Tang, A. (1995). Policy and General Technical Issues Related to Mitigating Seismic Effects on Electric Power and Communication Systems. In *Critical issues and state-of-the-art in lifeline earthquake engineering: proceedings of the session sponsored by the Technical Council on Lifeline Earthquake Engineering* (pp. 1-29).

Schot, J. (1998). Constructive Technology Assessment Comes of Age: The birth of a new politics of technology. In A. Jamison (Ed.), *Technology Policy Meets the Public (PESTO papers II)* (pp. 207-232): Aalborg University.

Schot, J. (2003). The Contested Rise of a Modernist Technology Politics. In T. Misa, P. Brey and A. Feenberg (Eds.), *Modernity and Technology* (pp. 257-278). Cambridge, Mass: The MIT Press.

Schot, J. and Rip, A. (1996). The Past and Future of Constructive Technology Assessment. *Technological Forecasting and Social Change*, 43(2,3), 251-268.

Sclove, R. E. (1996, July). Town Meetings on Technology. *Technology Review,* 24-31.

Sclove, R. E. (1999). Town Meetings on Technology. *Loka Institute.* Retrieved April, 2000. Available http://www.loka.org/pubs/techrev.htm

Shy, O. (2001). *The Economics of Network Industries.* New York: Cambridge University Press.

Sicker, D. C. (2002, Sept. 29). *Further Defining a Layered Model for Telecommunications Policy.* Paper presented at the Telecommunications Policy Research Conference, Alexandria, Virginia.

Sismondo, S. (1996). Exploring Metaphors of 'Social Construction'. In *Science Without Myth: On Constructions, Reality, and Social Knowledge* (pp. 49-78). Albany, NY: State University of New York Press.

Srivastava, L. and Samarajiva, R. (2001, June 27-29, 2001). *Regulatory design for disaster preparedness and recovery by infrastructure providers: South Asian experience.* Paper presented at the Fifth International Conference on Technology, Policy and Innovation, The Hague, Netherlands. Retrieved, from http://www.delft2001.tudelft.nl/

Strauss, A. and Corbin, J. (1990). *Basics of Qualitative Research: Grounded Theory Procedures and Techniques.* Newbury Park, CA: Sage.

Summerton, J. (1994a). Introductory Essay: The Systems Approach to Technological Change. In J. Summerton (Ed.), *Changing Large Technical Systems* (pp. 1-24). Boulder, CO: Westview Press.

Summerton, J. (Ed.). (1994b). *Changing Large Technical Systems.* Boulder, CO: Westview Press.

Takansashi, N., Tanaka, A., Yoshi, H. and Wada, Y. (1988). The Achilles' Heel of the Information Society: Socioeconomic Impacts of the Telecommunication Cable Fire in the Setagaya Telephone Office, Tokyo. *Technological Forecasting and Social Change*, 34, 27-52.

TELUS Communications (B.C.). (2000, Dec. 8). CRTC 8740-T46-4120/00 - TCBC TN 4120 - General Tariff - Wireless Service Provider Enhanced Provincial 9-1-1 Network Access Service.

TELUS Communications Inc. (2000, Dec. 20). CRTC 8740-T42 - TELUS Communications Inc. (TCI) - Tariff Applications for 2000- CRTC Tariff Notice 327: Wireless Service Provider Enhanced Provincial 9-1-1 Network Access Service.

TELUS Communications Inc. (2001a, Apr. 12). CRTC 8669-M16-01/01 - Microcell Telecommunications Inc. - Applications for Mandated Provision of Wireless Enhanced 9-1-1 Network Access Service - 2001/04/12 - TELUS Communications Inc.

TELUS Communications Inc. (2001b, Jan. 22). CRTC 8740-T42-327/00 - TCI TN 327 and TN 327/A - General Tariff - Wireless Service Provider Enhanced Provincial 9-1-1 Network Access Service - 2001/01/22 - TELUS Communications Inc.

TELUS Corporation. (2000, June 2). CRTC 8740-M29-0002/00 - Microcell Connexions - TN2 - TELUS Comments. Available http://www.crtc.gc.ca/8740/eng/2000/m29-2.htm

Thomas, B. (1994). *Emergency communications preparedness in Canada: a study of the command-and-control model and the emergence of alternative approaches.* Ottawa: National Library of Canada.

TSI Telecommunication Services Inc. (2002, Jan. 24). The Role of the 3rd Party Provider in Wireless 9-1-1. APCO/NENA Wireless E9-1-1 Implementation Plan Forum. Available http://www.nena9-1-1.org

United States Federal Communications Commission. (2001, Jan.). FCC Wireless 911 Requirements. Available http://www.fcc.gov

United States Federal Communications Commission. (2003, Nov. 13). Wireless E911 Coordination Initiative. Available http://wireless.fcc.gov/outreach/e911/

United States. (2003, Feb.). The National Strategy for the Physical Protection of Critical Infrastructures and Key Assets. Department of Homeland Security. Retrieved May, 2004. Available http://www.dhs.gov

Van Den Ende, J., Mulder, K., Knot, M., Moors, E. and Vergragt, P. (1998). Traditional and Modern Technology Assessment: Toward a Toolkit. *Technological Forecasting and Social Change*, 58(1,2), 5-21.

Webb, G. R., Tierney, K. and Dahlhamer, J. (2000). Businesses and Disasters: Empirical Patterns and Unanswered Questions. *Natural Hazards Review*, 1(2), 83-90.

Werbach, K. (2000, Sept. 1). *A Layered Model for Internet Policy.* Paper presented at the Telecommunications Policy Research Conference, Alexandria, Virginia.

Werle, R. (1998, Oct.). *The Internet: a new pattern of LTS development?* Paper presented at the EAAST 98 Conference, Session S14, 'The Political Sociology of Large Technical Systems', Atlanta, GA.

Winner, L. (1993). Upon Opening the Black Box and Finding it Empty: Social Constructivism and the Philosophy of Technology. *Science, Technology, and Human Values*, 18(3), 362-378.

Winner, L. (1995). Political Ergonomics. In R. Buchanan and V. Margolin (Eds.), *Discovering Design: Explorations in Design Studies*. Chicago: University of Chicago Press.

Winseck, D. (1998). *Reconvergence: A Political Economy of Telecommunications in Canada.* Cresskill, NJ: Hampton Press.

Wireless Location Industry Association. (2001). WLIA Online. Retrieved April, 2002. Available http://www.wliaonline.com

Wood, F. B. (1997). Lessons in Technology Assessment: Methodology and Management in OTA. *Technological Forecasting and Social Change*, 54(2,3), 145-162.

Wrobel, L. A. (1993). *Writing Disaster Recovery Plans for Telecommunications Networks and LANs*. Boston: Artech House.

Wrobel, L. A. (1997). *The Definitive Guide to Business Resumption Planning*. Boston: Artech House.

Index